T0234428

Robust Regression

STATISTICS: Textbooks and Monographs

A Series Edited by

Vol. 1: The Generalized Jackknife Statistic, *H. L. Gray and W. R. Schucany*
Vol. 2: Multivariate Analysis, *Anant M. Kshirsagar*
Vol. 3: Statistics and Society, *Walter T. Federer*
Vol. 4: Multivariate Analysis: A Selected and Abstracted Bibliography, 1957-1972, *Kocherlakota Subrahmaniam and Kathleen Subrahmaniam* (out of print)
Vol. 5: Design of Experiments: A Realistic Approach, *Virgil L. Anderson and Robert A. McLean*
Vol. 6: Statistical and Mathematical Aspects of Pollution Problems, *John W. Pratt*
Vol. 7: Introduction to Probability and Statistics (in two parts), Part I: Probability; Part II: Statistics, *Narayan C. Giri*
Vol. 8: Statistical Theory of the Analysis of Experimental Designs, *J. Ogawa*
Vol. 9: Statistical Techniques in Simulation (in two parts), *Jack P. C. Kleijnen*
Vol. 10: Data Quality Control and Editing, *Joseph I. Naus* (out of print)
Vol. 11: Cost of Living Index Numbers: Practice, Precision, and Theory, *Kali S. Banerjee*
Vol. 12: Weighing Designs: For Chemistry, Medicine, Economics, Operations Research, Statistics, *Kali S. Banerjee*
Vol. 13: The Search for Oil: Some Statistical Methods and Techniques, *edited by D. B. Owen*
Vol. 14: Sample Size Choice: Charts for Experiments with Linear Models, *Robert E. Odeh and Martin Fox*
Vol. 15: Statistical Methods for Engineers and Scientists, *Robert M. Bethea, Benjamin S. Duran, and Thomas L. Boullion*
Vol. 16: Statistical Quality Control Methods, *Irving W. Burr*
Vol. 17: On the History of Statistics and Probability, *edited by D. B. Owen*
Vol. 18: Econometrics, *Peter Schmidt*
Vol. 19: Sufficient Statistics: Selected Contributions, *Vasant S. Huzurbazar (edited by Anant M. Kshirsagar)*
Vol. 20: Handbook of Statistical Distributions, *Jagdish K. Patel, C. H. Kapadia, and D. B. Owen*
Vol. 21: Case Studies in Sample Design, *A. C. Rosander*
Vol. 22: Pocket Book of Statistical Tables, *compiled by R. E. Odeh, D. B. Owen, Z. W. Birnbaum, and L. Fisher*
Vol. 23: The Information in Contingency Tables, *D. V. Gokhale and Solomon Kullback*

Vol. 24: Statistical Analysis of Reliability and Life-Testing Models: Theory and Methods, *Lee J. Bain*
Vol. 25: Elementary Statistical Quality Control, *Irving W. Burr*
Vol. 26: An Introduction to Probability and Statistics Using BASIC, *Richard A. Groeneveld*
Vol. 27: Basic Applied Statistics, *B. L. Raktoe and J. J. Hubert*
Vol. 28: A Primer in Probability, *Kathleen Subrahmaniam*
Vol. 29: Random Processes: A First Look, *R. Syski*
Vol. 30: Regression Methods: A Tool for Data Analysis, *Rudolf J. Freund and Paul D. Minton*
Vol. 31: Randomization Tests, *Eugene S. Edgington*
Vol. 32: Tables for Normal Tolerance Limits, Sampling Plans, and Screening, *Robert E. Odeh and D. B. Owen*
Vol. 33: Statistical Computing, *William J. Kennedy, Jr. and James E. Gentle*
Vol. 34: Regression Analysis and Its Application: A Data-Oriented Approach, *Richard F. Gunst and Robert L. Mason*
Vol. 35: Scientific Strategies to Save Your Life, *I. D. J. Bross*
Vol. 36: Statistics in the Pharmaceutical Industry, *edited by C. Ralph Buncher and Jia-Yeong Tsay*
Vol. 37: Sampling from a Finite Population, *J. Hajek*
Vol. 38: Statistical Modeling Techniques, *S. S. Shapiro*
Vol. 39: Statistical Theory and Inference in Research, *T. A. Bancroft and C.-P. Han*
Vol. 40: Handbook of the Normal Distribution, *Jagdish K. Patel and Campbell B. Read*
Vol. 41: Recent Advances in Regression Methods, *Hrishikesh D. Vinod and Aman Ullah*
Vol. 42: Acceptance Sampling in Quality Control, *Edward G. Schilling*
Vol. 43: The Randomized Clinical Trial and Therapeutic Decisions, *edited by Niels Tygstrup, John M. Lachin, and Erik Juhl*
Vol. 44: Regression Analysis of Survival Data in Cancer Chemotherapy, *Walter H. Carter, Jr., Galen L. Wampler, and Donald M. Stablein*
Vol. 45: A Course in Linear Models, *Anant M. Kshirsagar*
Vol. 46: Clinical Trials: Issues and Approaches, *edited by Stanley H. Shapiro and Thomas H. Louis*
Vol. 47: Statistical Analysis of DNA Sequence Data, *edited by B. S. Weir*
Vol. 48: Nonlinear Regression Modeling: A Unified Practical Approach, *David A. Ratkowsky*
Vol. 49: Attribute Sampling Plans, Tables of Tests and Confidence Limits for Proportions, *Robert E. Odeh and D. B. Owen*
Vol. 50: Experimental Design, Statistical Models, and Genetic Statistics, *edited by Klaus Hinkelmann*
Vol. 51: Statistical Methods for Cancer Studies, *edited by Richard G. Cornell*
Vol. 52: Practical Statistical Sampling for Auditors, *Arthur J. Wilburn*
Vol. 53: Statistical Signal Processing, *edited by Edward J. Wegman and James G. Smith*
Vol. 54: Self-Organizing Methods in Modeling: GMDH Type Algorithms, *edited by Stanley J. Farlow*
Vol. 55: Applied Factorial and Fractional Designs, *Robert A. McLean and Virgil L. Anderson*
Vol. 56: Design of Experiments: Ranking and Selection, *edited by Thomas J. Santner and Ajit C. Tamhane*
Vol. 57: Statistical Methods for Engineers and Scientists. Second Edition, Revised and Expanded, *Robert M. Bethea, Benjamin S. Duran, and Thomas L. Boullion*
Vol. 58: Ensemble Modeling: Inference from Small-Scale Properties to Large-Scale Systems, *Alan E. Gelfand and Crayton C. Walker*
Vol. 59: Computer Modeling for Business and Industry, *Bruce L. Bowerman and Richard T. O'Connell*
Vol. 60: Bayesian Analysis of Linear Models, *Lyle D. Broemeling*

Vol. 61: Methodological Issues for Health Care Surveys, *Brenda Cox and Steven Cohen*

Vol. 62: Applied Regression Analysis and Experimental Design, *Richard J. Brook and Gregory C. Arnold*

Vol. 63: Statpal: A Statistical Package for Microcomputers – PC-DOS Version for the IBM PC and Compatibles, *Bruce J. Chalmer and David G. Whitmore*

Vol. 64: Statpal: A Statistical Package for Microcomputers – Apple Version for the II, II+, and IIe, *David G. Whitmore and Bruce J. Chalmer*

Vol. 65: Nonparametric Statistical Inference, Second Edition, Revised and Expanded, *Jean Dickinson Gibbons*

Vol. 66: Design and Analysis of Experiments, *Roger G. Petersen*

Vol. 67: Statistical Methods for Pharmaceutical Research Planning, *Sten W. Bergman and John C. Gittins*

Vol. 68: Goodness-of-Fit Techniques, *edited by Ralph B. D'Agostino and Michael A. Stephens*

Vol. 69: Statistical Methods in Discrimination Litigation, *edited by D.H. Kaye and Mikel Aickin*

Vol. 70: Truncated and Censored Samples from Normal Populations, *Helmut Schneider*

Vol. 71: Robust Inference, *M. L. Tiku, W. Y. Tan, and N. Balakrishnan*

Vol. 72: Statistical Image Processing and Graphics, *edited by Edward J. Wegman and Douglas J. DePriest*

Vol. 73: Assignment Methods in Combinatorial Data Analysis, *Lawrence J. Hubert*

Vol. 74: Econometrics and Structural Change, *Lyle D. Broemeling and Hiroki Tsurumi*

Vol. 75: Multivariate Interpretation of Clinical Laboratory Data, *Adelin Albert and Eugene K. Harris*

Vol. 76: Statistical Tools for Simulation Practitioners, *Jack P. C. Kleijnen*

Vol. 77: Randomization Tests, Second Edition, *Eugene S. Edgington*

Vol. 78: A Folio of Distributions: A Collection of Theoretical Quantile-Quantile Plots, *Edward B. Fowlkes*

Vol. 79: Applied Categorical Data Analysis, *Daniel H. Freeman, Jr.*

Vol. 80: Seemingly Unrelated Regression Equations Models : Estimation and Inference, *Virendra K. Srivastava and David E. A. Giles*

Vol. 81: Response Surfaces: Designs and Analyses, *Andre I. Khuri and John A. Cornell*

Vol. 82: Nonlinear Parameter Estimation: An Integrated System in BASIC, *John C. Nash and Mary Walker-Smith*

Vol. 83: Cancer Modeling, *edited by James R. Thompson and Barry W. Brown*

Vol. 84: Mixture Models: Inference and Applications to Clustering, *Geoffrey J. McLachlan and Kaye E. Basford*

Vol. 85: Randomized Response: Theory and Techniques, *Arijit Chaudhuri and Rahul Mukerjee*

Vol. 86: Biopharmaceutical Statistics for Drug Development, *edited by Karl E. Peace*

Vol. 87: Parts per Million Values for Estimating Quality Levels, *Robert E. Odeh and D. B. Owen*

Vol. 88: Lognormal Distributions: Theory and Applications, *edited by Edwin L. Crow and Kunio Shimizu*

Vol. 89: Properties of Estimators for the Gamma Distribution, *K. O. Bowman and L. R. Shenton*

Vol. 90: Spline Smoothing and Nonparametric Regression, *Randall L. Eubank*

Vol. 91: Linear Least Squares Computations, *R. W. Farebrother*

Vol. 92: Exploring Statistics, *Damaraju Raghavarao*

Vol. 93: Applied Time Series Analysis for Business and Economic Forecasting, *Sufi M. Nazem*

Vol. 94: Bayesian Analysis of Time Series and Dynamic Models, *edited by James C. Spall*

Vol. 95: The Inverse Gaussian Distribution: Theory, Methodology, and Applications, *Raj S. Chhikara and J. Leroy Folks*

Vol. 96: Parameter Estimation in Reliability and Life Span Models, *A. Clifford Cohen and Betty Jones Whitten*

Vol. 97: Pooled Cross-Sectional and Time Series Data Analysis, *Terry E. Dielman*

Vol. 98: Random Processes: A First Look, Second Edition, Revised and Expanded, *R. Syski*

Vol. 99: Generalized Poisson Distributions: Properties and Applications, *P.C. Consul*

Vol. 100: Nonlinear L_p-Norm Estimation, *René Gonin and Arthur H. Money*

Vol. 101: Model Discrimination for Nonlinear Regression Models, *Dale S. Borowiak*

Vol. 102: Applied Regression Analysis in Econometrics, *Howard E. Doran*

Vol. 103: Continued Fractions in Statistical Applications, *K.O. Bowman and L.R. Shenton*

Vol. 104: Statistical Methodology in the Pharmaceutical Sciences, *Donald A. Berry*

Vol. 105: Experimental Design in Biotechnology, *Perry D. Haaland*

Vol. 106: Statistical Methods in Pharmaceutical Drug Development, *edited by Karl E. Peace*

Vol. 107: Handbook of Nonlinear Regression Models, *David A. Ratkowsky*

Vol. 108: Robust Regression: Analysis and Applications, *edited by Kenneth D. Lawrence and and Jeffrey L. Arthur*

ADDITIONAL VOLUMES IN PREPARATION

Robust Regression

Analysis and Applications

edited by

Kenneth D. Lawrence

Rutgers University
Piscataway, New Jersey

Jeffrey L. Arthur

Oregon State University
Corvallis, Oregon

CRC Press
Taylor & Francis Group
Boca Raton London New York

CRC Press is an imprint of the
Taylor & Francis Group, an **informa** business

First published 1990 by Marcel Dekker, Inc.

Published 2019 by CRC Press
Taylor & Francis Group
6000 Broken Sound Parkway NW, Suite 300
Boca Raton, FL 33487-2742

First issued in paperback 2020

ISBN 13: 978-0-367-58018-6 (pbk)
ISBN 13: 978-0-8247-8129-3 (hbk)

Visit the Taylor & Francis Web site at
http://www.taylorandfrancis.com

and the CRC Press Web site at
http://www.crcpress.com

Library of Congress Cataloging-in-Publication Data

Robust regression : analysis and applications / edited by Kenneth D. Lawrence, Jeffrey L. Arthur.
p. cm. — (Statistics, textbooks and monographs)
Includes bibliographical references.
ISBN 0-8247-8129-5 (alk. paper)
1. Regression analysis. I. Lawrence, Kenneth D. II. Arthur, Jeffrey L. III. Series
QA278.2.R59 1990
519.5'36--dc20 89-23760

Preface

Statistical inference deals with the extraction of information from observations. Of equal importance to the empirical data are the assumptions underlying the analysis. While it is granted that these assumptions concerning randomness, independence, and so forth are not precisely true in the real setting, they are nonetheless invoked in order to provide theoretical foundations for the ensuing analysis.

The implicit assumption made in many such situations is that small deviations from the assumed form of the model will result in only small errors in the final results. Recent studies have indicated, however, that this is not always the case. As a result, the use of more robust statistical procedures has been an area of increased interest, both for the theoretician and the applied statistician. Robust statistical procedures, to paraphrase Huber (1981), are those that are insensitive to small deviations from the assumptions.

The purpose of this book is to bring together within one cover a variety of recent results on robust regression analysis. This collection of papers has been divided into four parts. The first part, on advances in robust regression, begins with a chapter by Chen and Box in which they characterize robust estimators in terms of how much they weight each observation. The weighting pattern of a Bayesian robust estimator is explored and compared with some M-estimators. The chapter by

Sposito presents some generalized properties of L_p-estimators and uses these properties to develop efficient computational procedures for obtaining the estimators. Lawrence and Arthur conclude this section with a description of their experiences in using various robust regression techniques on several small data sets.

Part II is concerned with robust regression methods and begins with a chapter by Armstrong and Beck in which they develop an algorithm for identifying outliers when using a least absolute value criterion in regression modeling. Computational results with primal and dual variants of the algorithm are presented. Morgenthaler discusses the use of redescending M-estimators in instances in which regression, when used for data description, provides multiple solutions. In their chapter, Narula and Wellington study L_1 linear regression and provide sensitivity analyses on the intervals in which the response variable and regressor variable may lie and not affect the resulting regression line. The chapter by Fellner proposes the use of best linear unbiased estimates for estimating the fixed parameters and the random errors in the mixed linear model, and presents a robust method for estimating the variance components based on likelihood methods. Two data sets are analyzed to illustrate the concepts.

The third part is devoted to forecasting and robust regression. The section begins with the chapter by Cogger, which summarizes some of the known properties of L_1 estimates for time series analysis, describes how existing computer algorithms can be modified to obtain them, and compares an L_1 approach to the more common least squares approach on an actual time series data set. This is followed by the chapter by Lawrence, Hakak, and Gold, who use robust regression methods to estimate the parameters in a short-term forecasting function for economic analyses. Guerts and Tolley also concern themselves with time series data but focus on the problem of identifying outliers via a method that is simple to implement and does not require complete identification of the model parameters. The method is applied to Hawaii tourist data. Guerard and Lawrence examine the use of ordinary least squares, latent root regression, and a robust regression weighting scheme to analyze data on mergers in the United States during this century. Guerts, Rinne, and Lawrence investigate the use of regression models for forecasting sales and provide alternative ways of dealing with outliers in such models.

The final part of the book presents two papers on robust ridge regression. Pfaffenberger and Dielman report the results of an experiment comparing five different robust ridge regression estimators in those instances when both multicollinearity (correlation among the explanatory variables) and nonnormal error distributions are present. Finally, Guerard and Ochsner study composite earnings models and examine the issue of robust weighting in their estimation. Multicollinearity between analyst forecasts and time series forecasts suggests the use of ridge regression techniques.

An endeavor such as this could not be completed without the efforts of many individuals. We would like to thank Mr. Brian Black, Senior Production Editor at Marcel Dekker, for his endless assistance and patience in the technical preparation of this volume. Of course, without the contributors and reviewers, a project such as this would be impossible; their energy is hereby acknowledged. Finally, our appreciation goes to our departments and our colleagues at Rutgers University and Oregon State University for providing the atmosphere conducive to such tasks.

Kenneth D. Lawrence
Jeffrey L. Arthur

Contents

Preface iii

Contributors xi

Part I Advances in Robust Regression

1. The Weighting Pattern of a Bayesian Estimator 3
   Gina G. Chen and George P. Box

2. Some Properties of L_p-Estimators 23
   Vincent A. Sposito

3. Robust Nonlinear Regression 59
   Kenneth D. Lawrence and Jeffrey L. Arthur

Part II Robust Regression Methods

4. An Algorithm to Assist in the Identification of Multiple Multivariate Outliers When Using a Least Absolute Value Criterion 89

Ronald D. Armstrong and Philip O. Beck

5. Fitting Redescending M-Estimators in Regression 105

Stephan Morgenthaler

6. On the Robustness of the Simple Linear Minimum Sum of Absolute Errors Regression 129

Subhash C. Narula and John F. Wellington

7. Residuals in Variance-Component Models

William H. Fellner

Part III Forecasting and Robust Regression

8. Robust Time Series Analysis—An L_1 Approach 173

Kenneth O. Cogger

9. Robust Regression and Input-Output Forecasts 181

Sheila M. Lawrence, Ellie Hakak, and M. Gold

10. Using an Empirical Transformation Technique to Detect Outliers for Improved Accuracy in Forecasting Models 195

Michael D. Geurts and H. Dennis Tolley

11. A New Look at the Mergers in the United States, 1895–1973 213

John B. Guerard, Jr. and Kenneth D. Lawrence

12. Alternative Methods of Dealing with Outliers in Forecasting Sales with Regression-Based Models 225

Michael D. Geurts, Heikki J. Rinne, and Sheila M. Lawrence

Part IV Robust Ridge Regression

13. A Comparison of Regression Estimators When Both Multicollinearity and Outliers Are Present 243

Roger C. Pfaffenberger and Terry E. Dielman

14. Composite Earnings Forecasting Efficiency and Executive Compensation 271

John B. Guerard, Jr., and Robert Ochsner

Index 285

Contributors

RONALD D. ARMSTRONG Professor of Management Science, Graduate School of Management, Rutgers University, Newark, New Jersey

JEFFREY L. ARTHUR Associate Professor, Department of Statistics, Oregon State University, Corvallis, Oregon

PHILIP O. BECK Manager, Financial Analysis, AMR Eagle, American Airlines, Dallas, Texas

GEORGE P. BOX Director, Center for Quality and Productivity Improvement, Department of Statistics and Industrial Engineering, University of Wisconsin-Madison, Madison, Wisconsin

GINA G. CHEN Quality Engineering Manager, Quality Department, Hewlett-Packard, New Jersey Division, Rockaway, New Jersey

KENNETH O. COGGER Professor, School of Business, The University of Kansas, Lawrence, Kansas

TERRY E. DIELMAN Associate Professor, Decision Sciences, M. J. Neeley School of Business, Texas Christian University, Fort Worth, Texas

WILLIAM H. FELLNER Consultant, Engineering Department, E. I. du Pont de Nemours and Company, Inc., Wilmington, Delaware

MICHAEL D. GEURTS J. Darwin Gunnell Professor, Institute of Business Management, Brigham Young University, Provo, Utah

M. GOLD Department of Economics, New York University, New York, New York

JOHN B. GUERARD, Jr. Senior Quantitative Analyst, Drexel Burnham Lambert, Chicago, Illinois

ELLIE HAKAK Economic Consultant, ELEONOMIC, Flushing, New York

KENNETH D. LAWRENCE Professor, Department of Industrial and Systems Engineering, Rutgers University, Piscataway, New Jersey

SHEILA M. LAWRENCE Rutgers University, North Brunswick, New Jersey

STEPHAN MORGENTHALER* Associate Professor, Department of Statistics, Yale University, New Haven, Connecticut

SUBHASH C. NARULA Professor, Decision Sciences and Business Law, Virginia Commonwealth University, Richmond, Virginia

ROBERT OCHSNER Director of Compensation, Hay Management Consultants, Philadelphia, Pennsylvania

ROGER C. PFAFFENBERGER M-Bank Professor of Management Science, M. J. Neeley School of Business, Texas Christian University, Fort Worth, Texas

HEIKKI J. RINNE Associate Professor, Institute of Business Management, Brigham Young University, Provo, Utah

---

*Current affiliation: Professor, Department of Mathematics, Ecole Polytechnique Federal de Lausanne, Lausanne, Switzerland.

VINCENT A. SPOSITO Professor, Department of Statistics, Iowa State University, Ames, Iowa

H. DENNIS TOLLEY Professor, Department of Statistics, Brigham Young University, Provo, Utah

JOHN F. WELLINGTON Associate Professor of Management Science, MBA Program, Gannon University, Erie, Pennsylvania

Robust Regression

Part I

Advances in Robust Regression

1

The Weighting Pattern of a Bayesian Robust Estimator

GINA G. CHEN Hewlett-Packard, Rockaway, New Jersey

GEORGE P. BOX University of Wisconsin–Madison, Madison, Wisconsin

1. INTRODUCTION

Data has frequently been analyzed as if, to an adequate approximation, errors are normally, identically, and independently distributed. Because it has come to be believed that the first two of the assumptions are frequently inappropriate and in fact that error distributions are likely to be leptokurtic and/or contaminated by occasional bad values giving rise to outliers, attention has been directed, in particular by Huber (1972), Andrews et al. (1972), Hogg (1974), and Barnett and Lewis (1978), to various estimators which are insensitive to such departures.

Such procedures usually give smaller weights to observations that appear discrepant and so can be characterized by their weighting patterns.

In Sections 2 and 3 the structure of certain robust Bayesian estimators is examined from this point of view, and in Section 4 the weighting pattern for a robust Bayesian posterior mean is compared with those for some M-estimators.

2. A BAYESIAN APPROACH TO ROBUST ESTIMATION

Consider the linear model

$$Y = X\theta + e$$

where Y is an $n \times 1$ vector of observations, X an $n \times \ell$ matrix of fixed elements with rank ℓ, θ an $\ell \times 1$ vector of regression coefficients, and e an $n \times 1$ vector of random errors. Following Jeffreys (1932), Dixon (1953), and Tukey (1960), suppose that independently each of the errors is either $N(0,\sigma^2)$ with probability $(1 - \alpha)$ or $N(0, k^2\sigma^2)$ with probability α, that is, follows a mixture distribution with k fixed, and suppose that the θ's and $\log \sigma$ are independent and locally uniform a priori.

Let $a_{(b)}$ be the event that a particular set of b of the n e_i's are from $N(0, k^2\sigma^2)$ and the remaining $n - b$ are from $N(0,\sigma^2)$ and correspondingly let the vector e be partitioned into $e_{(b)}$ and $e_{(n-b)}$, the vector Y into $Y_{(b)}$ and $Y_{(n-b)}$, and the matrix X into $X_{(b)}$, $X_{(n-b)}$. With this contaminated normal model the posterior distribution can be expressed as a weighted average of 2^n distributions

$$P(\theta \mid Y) = \sum_{(b)} q_{(b)} P(\theta \mid a_{(b)}, Y)$$

where using results in Box and Tiao (1968)

$$q_{(b)} = P(a_{(b)} \mid Y) = C\left(\frac{\alpha}{1-\alpha}\right)^b k^{-b} \frac{|X'X|^{1/2}}{|X'X - \phi X'_{(b)}X_{(b)}|^{1/2}} \left\{\frac{s^2_{(b)}}{s^2}\right\}^{-\nu/2} \tag{1}$$

$$P(\theta \mid a_{(b)}, Y) = \frac{\Gamma\left(\frac{n}{2}\right)|X'X - \phi X'_{(b)}X_{(b)}|^{1/2}}{\Gamma\left(\frac{\nu}{2}\right)(\pi\nu s^2_{(b)})^{\ell/2}} \times \left\{1 + \frac{(\theta - \hat{\theta}_{(b)})(X'X - \phi X'_{(b)}X_{(b)})(\theta - \hat{\theta}_{(b)})}{\nu s^2_{(b)}}\right\}^{-n/2} \tag{2}$$

and

$$\hat{\theta}_{(b)} = \left(X'_{(n-b)}X_{(n-b)} + \frac{1}{k^2}X'_{(b)}X_{(b)}\right)^{-1}\left(X'_{(n-b)}Y_{(n-b)} + \frac{1}{k^2}X'_{(b)}Y_{(b)}\right) \tag{3}$$

$$s^2_{(b)} = \frac{1}{\nu} S_{(b)}(\hat{\theta}_{(b)})$$

$$= \frac{1}{\nu} \{(Y_{(n-b)} - X_{(n-b)}\hat{\theta}_{(b)})'(Y_{(n-b)} - X_{(n-b)}\hat{\theta}_{(b)})$$

$$+ \frac{1}{k^2} (Y_{(b)} - X_{(b)}\hat{\theta}_{(b)})'(Y_{(b)} - X_{(b)}\hat{\theta}_{(b)})\} \quad (4)$$

$$\nu = n - \ell, \quad \phi = 1 - k^{-2}, \quad s^2 = s_0^2$$

and C is the constant making the $q_{(b)}$'s sum to unity.

The posterior distribution $P(\theta|a_{(b)}, Y)$ is an ℓ-dimensional multivariate t-distribution with mean $\hat{\theta}_{(b)}$, dispersion matrix $s^2_{(b)}(X'X - \phi(X'_{(b)}X_{(b)}))^{-1}$, and $\nu = n - \ell$ degrees of freedom.

An estimator robust to this type of contamination which has the property of minimizing the mean square error loss is the posterior mean $\Sigma q_{(b)}\hat{\theta}_{(b)}$.

Another alternative estimator is the <u>posterior mode</u>, which is also the maximum likelihood estimator. Comparisons of the posterior mode with the M-estimators were discussed in Chen and Box (1979); the results are similar to those of the present study. In this chapter we will concentrate on the posterior mean.

3. THE WEIGHTING STRUCTURE OF THE POSTERIOR MEAN

3.1 The General Linear Model

To define how the posterior mean $\Sigma q_{(b)}\hat{\theta}_{(b)}$ weights each observation, consider the following two equations in Box and Tiao (1968):

$$\nu s^2_{(b)} = \nu s^2$$

$$- \phi(Y_{(b)} - X_{(b)}\hat{\theta})'\{I - \phi X_{(b)}(X'X)^{-1}X'_{(b)}\}^{-1}(Y_{(b)} - X_{(b)}\hat{\theta}) \quad (5)$$

$$\hat{\theta}_{(b)} = \hat{\theta} - \phi(X'X)^{-1}X'_{(b)}\{I - \phi X_{(b)}(X'X)^{-1}X'_{(b)}\}^{-1}(Y_{(b)} - X_{(b)}\hat{\theta}) \quad (6)$$

Using (6),

$$Y_{(b)} - X_{(b)}\hat{\theta}_{(b)} = Y_{(b)} - X_{(b)}\hat{\theta}$$
$$+ \phi X_{(b)}(X'X)^{-1}X'_{(b)}\{I - \phi X_{(b)}(X'X)^{-1}X'_{(b)}\}^{-1}$$
$$\times (Y_{(b)} - X_{(b)}\hat{\theta})$$
$$= [I + \phi X_{(b)}(X'X)^{-1}X'_{(b)}\{I - \phi X_{(b)}(X'X)^{-1}X'_{(b)}\}^{-1}]$$
$$\times (Y_{(b)} - X_{(b)}\hat{\theta})$$
$$= \{I - \phi X_{(b)}(X'X)^{-1}X'_{(b)}\}^{-1}(Y_{(b)} - X_{(b)}\hat{\theta}) \tag{7}$$

or equivalently,

$$Y_{(b)} - X_{(b)}\hat{\theta} = \{I - \phi X_{(b)}(X'X)^{-1}X'_{(b)}\}(Y_{(b)} - X_{(b)}\hat{\theta}_{(b)}) \tag{8}$$

Dividing both sides of equation (5) by $\nu s^2_{(b)}$ and substituting $Y_{(b)} - X_{(b)}\hat{\theta}$ by the right-hand side of equation (8), we have

$$\frac{s^2}{s^2_{(b)}} = 1 + \frac{\phi}{\nu}\left(\frac{Y_{(b)} - X_{(b)}\hat{\theta}_{(b)}}{s_{(b)}}\right)'$$
$$\times \{I - \phi X_{(b)}(X'X)^{-1}X'_{(b)}\}\left(\frac{Y_{(b)} - X_{(b)}\hat{\theta}_{(b)}}{s_{(b)}}\right) \tag{9}$$

Equation (1) can now be written as

$$q_{(b)} = C\left(\frac{\alpha}{1 - \alpha}\right)^b k^{-b} \frac{|X'X|^{1/2}}{|X'X - \phi X'_{(b)}X_{(b)}|^{1/2}}$$
$$\times \left[1 + \frac{\phi}{\nu}\left(\frac{Y_{(b)} - X_{(b)}\hat{\theta}_{(b)}}{s_{(b)}}\right)\{I - \phi X_{(b)}(X'X)^{-1}X'_{(b)}\}\right.$$
$$\left.\times \left(\frac{Y_{(b)} - X_{(b)}\hat{\theta}_{(b)}}{s_{(b)}}\right)\right]^{\nu/2} \tag{10}$$

Let

$$\Sigma = \phi(I - \phi X_{(b)}(X'X)^{-1}X'_{(b)})\left(\frac{\nu - b}{\nu}\right)$$

$$\tilde{r}_{(b)} = \frac{Y_{(b)} - X_{(b)}\hat{\theta}_{(b)}}{s_{(b)}}$$

Since

$$|X'X||\Sigma| = \begin{vmatrix} X'X & X'_{(b)} \\ \phi X_{(b)} & I \end{vmatrix} \left(\frac{\nu - b}{\nu}\phi\right)^b$$

$$= \begin{vmatrix} I & \phi X_{(b)} \\ X'_{(b)} & X'X \end{vmatrix} \left(\frac{\nu - b}{\nu}\phi\right)^b$$

$$= |X'X - \phi X'_{(b)}X_{(b)}| \left(\frac{\nu - b}{\nu}\phi\right)^b$$

equation (10) becomes

$$q_{(b)} = C\left(\frac{\alpha}{1 - \alpha}\right)^b k^{-b}\left(\frac{\nu - b}{\nu}\phi\right)^{b/2} |\Sigma|^{-1/2}\left(1 + \frac{1}{\nu - b}\tilde{r}'_{(b)}\Sigma\tilde{r}_{(b)}\right)^{\nu/2} \tag{11}$$

Thus $q_{(b)}$ is proportional to the inverse of a multivariate t ordinate at $\tilde{r}_{(b)}$ with precision matrix Σ and $\nu - b$ degrees of freedom.

Recall that $\tilde{r}_{(b)} = (Y_{(b)} - X_{(b)}\hat{\theta}_{(b)})/s_{(b)}$, where $\hat{\theta}_{(b)}$ and $s_{(b)}$ are estimates for the location and scale parameters, respectively, on the assumption that $a_{(b)}$ occurs. The vector $\tilde{r}_{(b)}$ is then the standardized residual vector of observations supposed to have come from the distribution with a larger variance, $N(0, k^2\sigma^2)$. It is seen that $\tilde{r}_{(b)}$ is an important factor determining the weighting pattern. If the standardized residuals are large, then $q_{(b)}$ will be large and $\hat{\theta}_{(b)}$ will be a more influential component of the posterior mean.

Notice that $\hat{\theta}_{(b)}$ is a weighted least square estimate with weights inversely proportional to the variance of each observation under the assumption $a_{(b)}$. These weighted least square estimates are again weighted based on the size of the standardized residuals under $a_{(b)}$ and averaged to produce the posterior mean. The weighting pattern is the single most important feature of this Bayesian estimator $\Sigma q_{(b)}\hat{\theta}_{(b)}$.

Most robust estimators can be characterized by their weighting patterns, although different criteria have been used in choosing the weights. In L-estimators, the weight given to an observation depends on the percentage point it takes in the ordered sample. In M-estimators, while it is often characterized by the influence function $\psi(x)$, the weighting function can easily be obtained by taking $w(x) = \psi(x)/x$, as is used in Tukey's biweight M-estimator. Moreover, the weights in an M-estimator are essentially determined through standardized residuals obtained iteratively. This is very similar to what we have seen in the Bayesian posterior mean where standardized residual vector $\tilde{r}_{(b)}$ determines $q_{(b)}$. It is natural to compare the weighting functions $w(x)$.

3.2 The Simple Location Model

Useful insight is gained from considering the estimate of a simple location parameter.

Consider the model $y_i = \theta + e_i$, where e_i's are from a contaminated normal distribution $(1 - \alpha)N(0, \sigma^2) + \alpha N(0, k^2\sigma^2)$. Then $X'X = n$, $X'_{(b)}X_{(b)} = b$, $I - \phi X_{(b)}(X'X)^{-1}X'_{(b)} = I - (\phi/n)X_{(b)}X'_{(b)}$, and following (11),

$$q_{(b)} = C\left(\frac{\alpha}{1 - \alpha}\right)^b k^{-b}\left(\frac{n}{n - b\phi}\right)^{1/2}$$

$$\times \left[1 + \frac{\phi}{n - 1}\tilde{r}'_{(b)}\left\{I - \frac{\phi}{n}X_{(b)}X'_{(b)}\right\}\tilde{r}_{(b)}\right]^{(n-1)/2} \tag{12}$$

Also, from (6)

$$\hat{\theta}_{(b)} = \bar{y} - \frac{b\phi}{n - b\phi}(\bar{y}_{(b)} - \bar{y})$$

where $\bar{y}$ is the sample average and $\bar{y}_{(b)}$ is the average of elements of $Y_{(b)}$.

The posterior mean $\Sigma_{(b)} q_{(b)}\hat{\theta}_{(b)}$ can alternatively be written as $\Sigma_{b=0}^{n} P(b|\underset{\sim}{y}, \alpha)\hat{\theta}^{(b)}$ where $P(b|\underset{\sim}{y}, \alpha)$ is the posterior probability that there are b bad values and $\hat{\theta}^{(b)}$ is the corresponding conditional posterior mean so that $\hat{\theta}^{(0)}$ is the usual least squares estimate. In particular, for this location model we can write the posterior mean as $\Sigma_{b=0}^{n} P(b|\underset{\sim}{y}, \alpha)y^{(b)}$ with $y^{(0)} = y$ (the sample mean) and

$$\bar{y}^{(1)} = \sum_{i=1}^{n} q_i\hat{\theta}_{(i)} \Big/ \sum_{i=1}^{n} q_i$$

$$\bar{y}^{(2)} = \sum_{\substack{i,j=1 \\ i<j}}^{n} q_{i,j}\hat{\theta}_{(i,j)} \Big/ \sum_{\substack{i,j=1 \\ i<j}}^{n} q_{i,j}$$

$\vdots$

etc.

where q_i is the posterior probability that the ith observation is from $N(\theta, k^2\sigma^2)$, and $q_{i,j}$ is the posterior probability that the ith and the jth observations are from $N(\theta, k^2\sigma^2)$. From (12)

$$q_i = C\left(\frac{\alpha}{1-\alpha}\right)k^{-1}\left(\frac{n}{n-\phi}\right)^{1/2}\left(1 + \frac{\phi}{n-1}\,\frac{n-\phi}{n}\,\tilde{r}_i^2\right)^{(n-1)/2} \tag{13}$$

Let

$$D_i = \left(1 + \frac{\phi}{n-1}\,\frac{n-\phi}{n}\,\tilde{r}_i^2\right)^{(n-1)/2} \quad \text{and} \quad D = \sum_{i=1}^{n} D_i$$

then

$$\bar{y}^{(1)} = \sum_{i=1}^{n} \frac{D - \phi D_i}{(n-\phi)D} y_i \tag{14}$$

Similarly,

$$q_{i,j} = C\left(\frac{\alpha}{1-\alpha}\right)^2 k^{-2}\left(\frac{n}{n-2\phi}\right)^{1/2}\left\{1 + \frac{\phi}{n-1}\,\tilde{r}'_{i,j}\begin{pmatrix} \frac{n-\phi}{n} & -\frac{\phi}{n} \\ -\frac{\phi}{n} & \frac{n-\phi}{n} \end{pmatrix}\tilde{r}_{i,j}\right\}^{(n-1)/2}$$

Let

$$T_{ij} = \left\{1 + \frac{\phi}{n-1}\,\tilde{r}'_{i,j}\begin{pmatrix} \frac{n-\phi}{n} & -\frac{\phi}{n} \\ -\frac{\phi}{n} & \frac{n-\phi}{n} \end{pmatrix}\tilde{r}_{i,j}\right\}^{(n-1)/2}$$

and

$$T = \frac{1}{2} \sum_{\substack{i,j \\ i \neq j}} T_{ij}$$

then

$$\bar{y}^{(2)} = \sum_{i=1}^{n} \frac{\left(T - \phi \sum_{\substack{j=1 \\ j \neq i}}^{n} T_{ij} \right)}{(n - 2\phi)T} y_i \qquad (15)$$

The formulas for $\bar{y}^{(3)}$, $\bar{y}^{(4)}$, . . ., etc. are obtained in the same fashion.

The posterior probabilities $P(b \mid \underset{\sim}{y}, \alpha)$ are

$$P(0 \mid \alpha) = q_0, \quad p(1 \mid \alpha) = \sum_{i=1}^{n} q_i, \quad P(2 \mid \alpha) = \sum_{\substack{i,j=1 \\ i<j}}^{n} q_{i,j}, \ldots, \text{ etc.}$$

Consider the case where the probabilities for three or more bad values are negligible; then

$$P(0 \mid \underset{\sim}{y}, \alpha) = \frac{1}{1 + \epsilon\left(\frac{n}{n-\phi}\right)^{1/2} D + \epsilon^2 \left(\frac{n}{n-2\phi}\right)^{1/2} T}$$

$$P(1 \mid \underset{\sim}{y}, \alpha) = \frac{\epsilon\left(\frac{n}{n-\phi}\right)^{1/2} D}{1 + \epsilon\left(\frac{n}{n-\phi}\right)^{1/2} D + \epsilon^2 \left(\frac{n}{n-2\phi}\right)^{1/2} T}$$

$$P(2 \mid \underset{\sim}{y}, \alpha) = \frac{\epsilon^2\left(\frac{n}{n-2\phi}\right)^{1/2} T}{1 + \epsilon\left(\frac{n}{n-\phi}\right)^{1/2} D + \epsilon^2 \left(\frac{n}{n-2\phi}\right)^{1/2} T}$$

$$\text{where } \epsilon = \frac{\alpha}{1-\alpha} \frac{1}{k} \qquad (16)$$

Notice that the posterior means depend on α only through the $P(b \mid \underset{\sim}{y}, \alpha)$ so that the role that α plays, as expected, is only to adjust the probability associated with fixed numbers of bad values. When k is fairly large, the quantity $\phi = 1 - 1/k^2$ is close to one ($\phi = .96$ when k = 5) and $\bar{y}^{(1)}$, $\bar{y}^{(2)}$ are insensitive to k since they involve only ϕ, as are D and T. If we let

$$\epsilon = \frac{\alpha}{1-\alpha}\frac{1}{k}$$

and approximate ϕ by one, then the posterior mean is mainly a function of ϵ.

This quantity ϵ measures the ratio of information about θ coming from $N(\theta, k^2\sigma^2)$ to that coming from $N(\theta, \sigma^2)$. Thus ϵ is equal to

$$\frac{\begin{pmatrix}\text{Probability the observation}\\ y \text{ is from } N(\theta, k^2\sigma^2)\end{pmatrix} \times \begin{pmatrix}\text{Fisher information about } \theta\\ \text{contained in } y \text{ given } y \sim N(\theta, k^2\sigma^2)\end{pmatrix}^{1/2}}{\begin{pmatrix}\text{Probability the observation}\\ y \text{ is from } N(\theta, \sigma^2)\end{pmatrix} \times \begin{pmatrix}\text{Fisher information about } \theta\\ \text{contained in } y \text{ given } y \sim N(\theta, \sigma^2)\end{pmatrix}^{1/2}}$$

$$= \frac{\alpha\sqrt{1/k^2\sigma^2}}{(1-\alpha)\sqrt{1/\sigma^2}} = \frac{\alpha}{1-\alpha}\frac{1}{k} \tag{17}$$

For large k, therefore, we need to postulate a priori only one parameter in the model, namely ϵ. For smaller k, such as k < 5, the influence of k is not negligible. With fixed ϵ, the smaller k is, the more weight the posterior mean will put on observations with large standardized residuals, as one is not expecting drastic outliers in that case. It will be seen, however, that most of the proposed M-estimators really have weighting functions closer to posterior means with $k \geqslant 5$.

4. COMPARISONS OF THE BAYESIAN POSTERIOR MEAN AND SOME M-ESTIMATORS

As seen in Section 3.2, when a contaminated model is used, the Bayesian posterior mean may be written

$$\hat{\theta} = \sum_{b=0}^{n} P(b \mid \underset{\sim}{y}, \alpha)\hat{\theta}^{(b)} \tag{18}$$

Much insight can be obtained by studying the special case of a location model with the posterior probabilities $P(b \mid \underset{\sim}{y}, \alpha)$ negligible for $b \geq 2$. Using (14) and (16), (18) becomes

$$\hat{\theta} = P(0 \mid \underset{\sim}{y}, \alpha)\bar{y} + P(1 \mid \underset{\sim}{y}, \alpha)\bar{y}^{(1)}$$

$$= \frac{1}{1 + \epsilon\left(\frac{n}{n-\phi}\right)^{1/2} D} \frac{1}{n}\sum_{i=1}^{n} y_i + \frac{\epsilon\left(\frac{n}{n-2\phi}\right)^{1/2} D}{1 + \epsilon\left(\frac{n}{n-\phi}\right)^{1/2} D} \sum_{i=1}^{n} \frac{D - \phi D_i}{(n-\phi)D} y_i$$

$$= \sum_{i=1}^{n} \left\{ \frac{1}{n} \frac{1}{1 + \left(\frac{n}{n-\phi}\right)^{1/2} D} + \frac{\epsilon\left(\frac{n}{n-2\phi}\right)^{1/2} D}{1 + \epsilon\left(\frac{n}{n-\phi}\right)^{1/2} D} \frac{D - \phi D_i}{(n-\phi)D} \right\} y_i$$

$$= \sum_{i=1}^{n} W_i y_i \qquad (19)$$

The quantity

$$W_i = \frac{1}{n} \frac{1}{1 + \epsilon\left(\frac{n}{n-\phi}\right)^{1/2} D} + \frac{\epsilon\left(\frac{n}{n-2\phi}\right)^{1/2} D}{1 + \epsilon\left(\frac{n}{n-\phi}\right)^{1/2} D} \frac{D - \phi D_i}{(n-\phi)D}$$

is the weight the Bayesian posterior mean puts on the ith observation. It is clear that W_i depends on two factors: (1) the prior choice of ϵ (or α and k) and (2) the calculated value of D_i and D, which is determined by the value of y_i relative to the whole sample $\underset{\sim}{y}$. As was shown before,

$$D_i = \left(1 + \frac{\phi}{n-1} \frac{n-\phi}{n} \tilde{r}_i^2\right)^{(n-1)/2}$$

and the weight W_i is therefore a function of $\tilde{r}_i$. For convenient computation D_i can also be written as

$$D_i = \left(1 - \frac{n\phi}{n-\phi} r_i^2\right)^{-(n-1)/2} \qquad \text{with } r_i = \frac{y_i - \bar{y}}{s}$$

Notice that no matter how discordant an observation y_i may be, the weight W_i will never be zero; instead, W_i approaches a small number $(1 - \phi)/(n - \phi)$ as D_i approaches infinity. This is the result of assuming the discordant observations are from a normal distribution with the same mean and a larger variance.

Figure 1 shows plots of $W_1 - (1 - \phi)/(n - \phi)$ versus $\tilde{r}_1$ for three random normal samples of 10 observations with different ϵ (α and k) values when a multiple $m = 0, 1, 2, \ldots,$ of σ is added to the first observation in each sample. Although these Bayesian weights are sample adaptive, they remain remarkable stable in the sense that

1. The weight, after the discrepant observation is taken account of, is remarkably evenly spread throughout the remaining observations.
2. The weight points from different samples closely follow the same weighting curve.

The weighting patterns of the Bayesian posterior mean clearly depend heavily on the a priori choices of ϵ. When ϵ is very small the weights are nearly constant in the center and then decrease slowly to zero. As ϵ increases, the flat area in the center gets smaller and smaller. In all cases, the weighting patterns are like Huber's in the center and like Tukey's biweight in the tail. For readers who are used to thinking in these terms, an approximation to the influence curve is produced by plotting $W_1\tilde{r}_1$ against $\tilde{r}_1$. Figure 2 shows the "pseudo-influence curves" so obtained from Figure 1.

The Bayesian weights differ from those of most M-estimators in that they give small finite weight $(1 - \phi)/(n - \phi)$ even to highly discordant observations. For example, when $n = 10$, $k = 5.0$, then $(1 - \phi)/(n - \phi) = .004$; and when $n = 10$, $k = 10.0$, then $(1 - \phi)/(n - \phi) = .001$. This small nonzero weight, if not subtracted out, would result in a slow but steady increase in the tails of the pseudo-influence curves. These curves actually go to infinity with slope $(1 - \phi)/(n - \phi)$ although over practical ranges they remain small in the tail.

This may appear to some people as an undesirable property of this Bayesian posterior mean. However, the implication is usually that this small weight rather than zero weight should be given to outlying observations because they still could contain relevant information regarding the parameter to be estimated.

When $(1 - \phi)/(n - \phi)$ is subtracted from W_i's, the weighting curves can be approximated by the exponential power function $.1e^{-.00061|\tilde{r}_1|^b}$ as shown in Figure 1. The parameter b depends approximately linearly on log ϵ as shown in Figure 3. As an approximation, one can therefore calculate the Bayesian posterior mean using the weighting function $(1 - \phi)/(n - \phi) + .1e^{-.00061|\tilde{r}_i|^b}$.

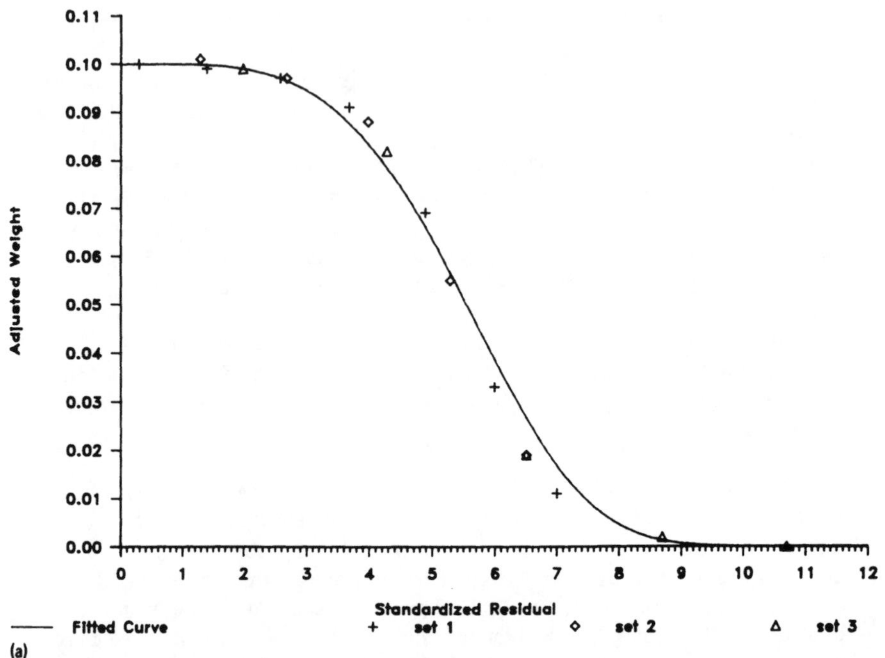

FIGURE 1. The weighting curves of Bayesian posterior means with different ϵ (α and k) values. (a) α = .01, k = 10.0, ϵ = .001, fitted curve = $.1e^{-.00061|\tilde{r}_1|^{4.1}}$. (b) α = .01, k = 5.0, ϵ = .002, fitted curve = $.1e^{-.00061|\tilde{r}_1|^{4.3}}$. (c) α = .05, k = 10.0, ϵ = .005, fitted curve = $.1e^{-.00061|\tilde{r}_1|^{4.6}}$. (d) α = .05, k = 5.0, ϵ = .011, fitted curve = $.1e^{-.00061|\tilde{r}_1|^{4.9}}$. (e) α = .10, k = 5.0, ϵ = .022, fitted curve = $.1e^{-.00061|\tilde{r}_1|^{5.2}}$.

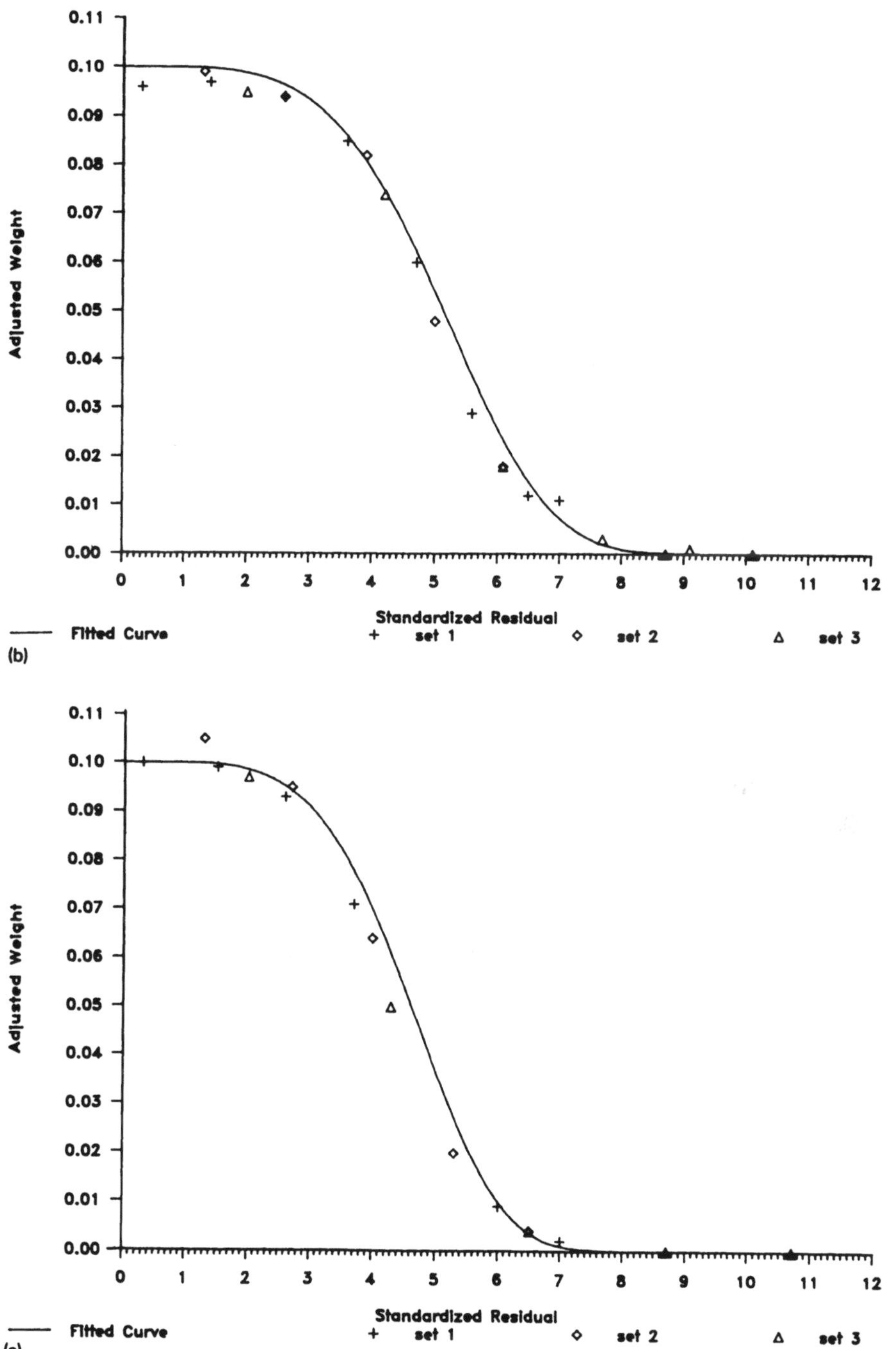
0.11
0.10
0.09
0.08
0.07
0.06
0.05
0.04
0.03
0.02
0.01
0.00
Adjusted Weight
0
1
2
3
4
5
6
7
8
9
10
11
12
Standardized Residual
Fitted Curve
set 1
set 2
set 3
(b)
0.11
0.10
0.09
0.08
0.07
0.06
0.05
0.04
0.03
0.02
0.01
0.00
Adjusted Weight
0
1
2
3
4
5
6
7
8
9
10
11
12
Standardized Residual
Fitted Curve
set 1
set 2
set 3
(c)

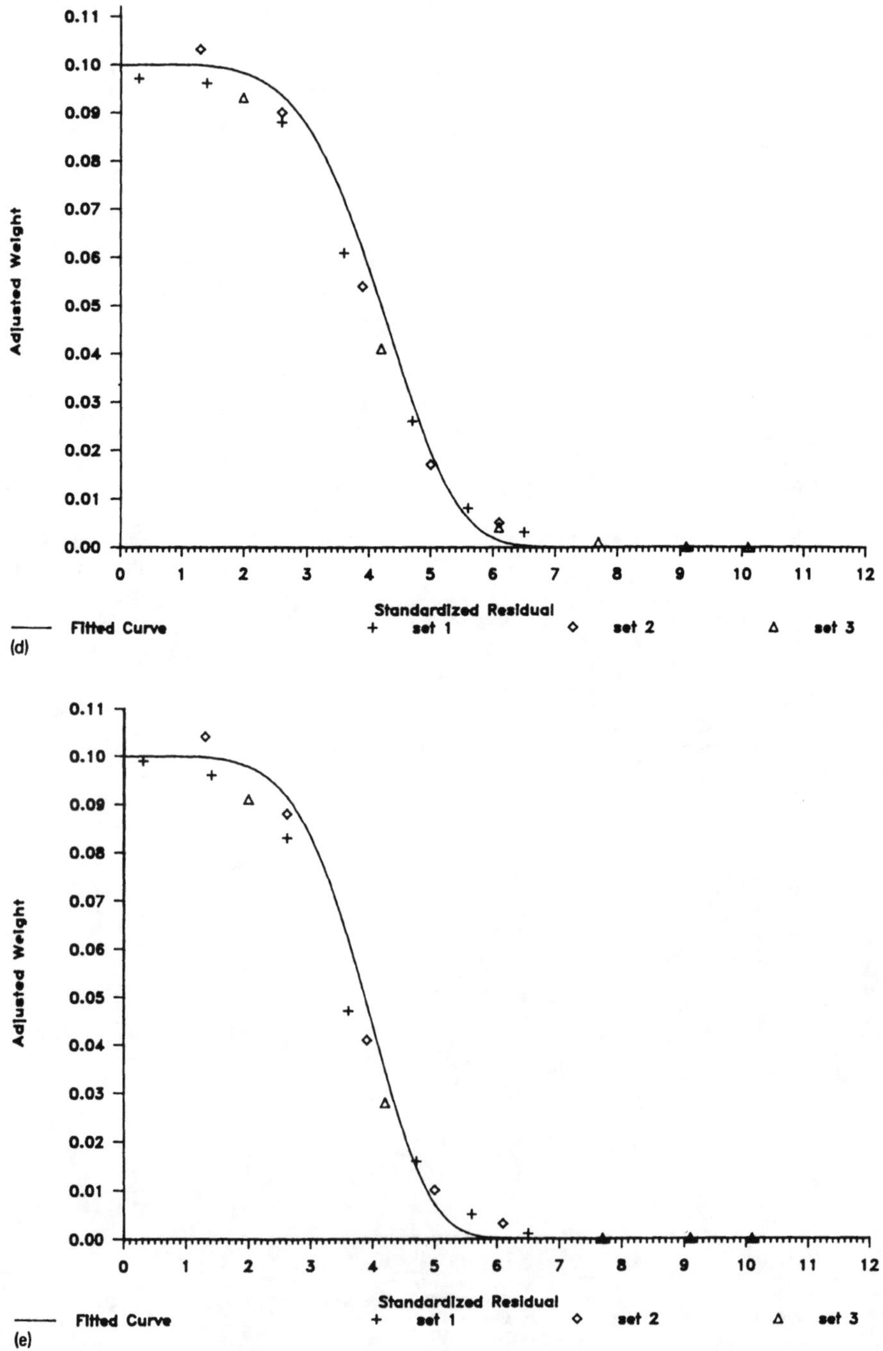

FIGURE 1 (Continued)

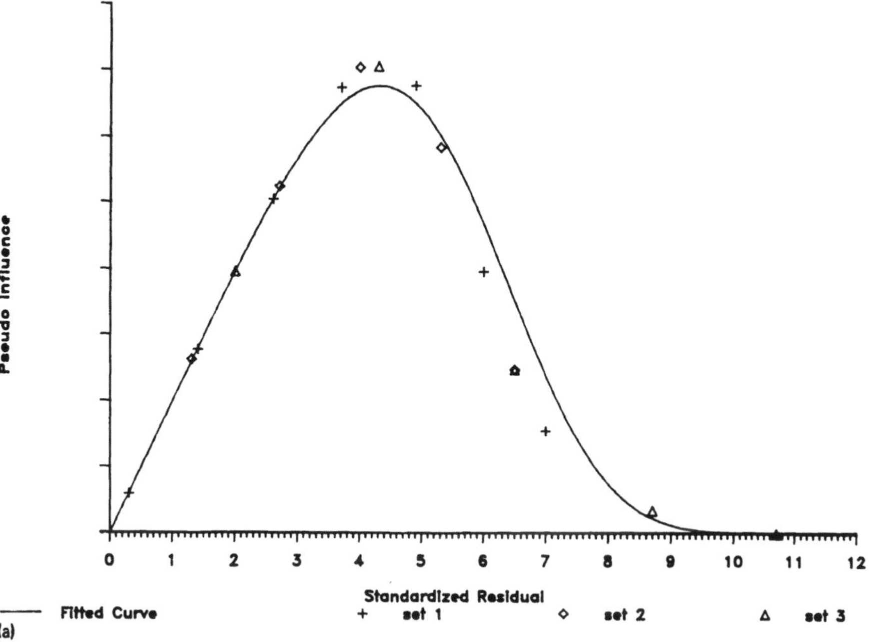

FIGURE 2 Pseudo-influence curves obtained from data and curves in Figure 1. (a) α = .01, k = 10.0, ϵ = .001, fitted curve = $.1\tilde{r}_1 e^{-.00061|\tilde{r}_1|^{4.1}}$.
(b) α = .01, k = 5.0, ϵ = .002, fitted curve = $.1\tilde{r}_1 e^{-.00061|\tilde{r}_1|^{4.3}}$.
(c) α = .05, k = 10.0, ϵ = .005, fitted curve = $.1\tilde{r}_1 e^{-.00061|\tilde{r}_1|^{4.6}}$.
(d) α = .05, k = 5.0, ϵ = .011, fitted curve = $.1\tilde{r}_1 e^{-.00061|\tilde{r}_1|^{4.9}}$.
(e) α = .10, k = 5.0, ϵ = .022, fitted curve = $.1\tilde{r}_1 e^{-.00061|\tilde{r}_1|^{5.2}}$.

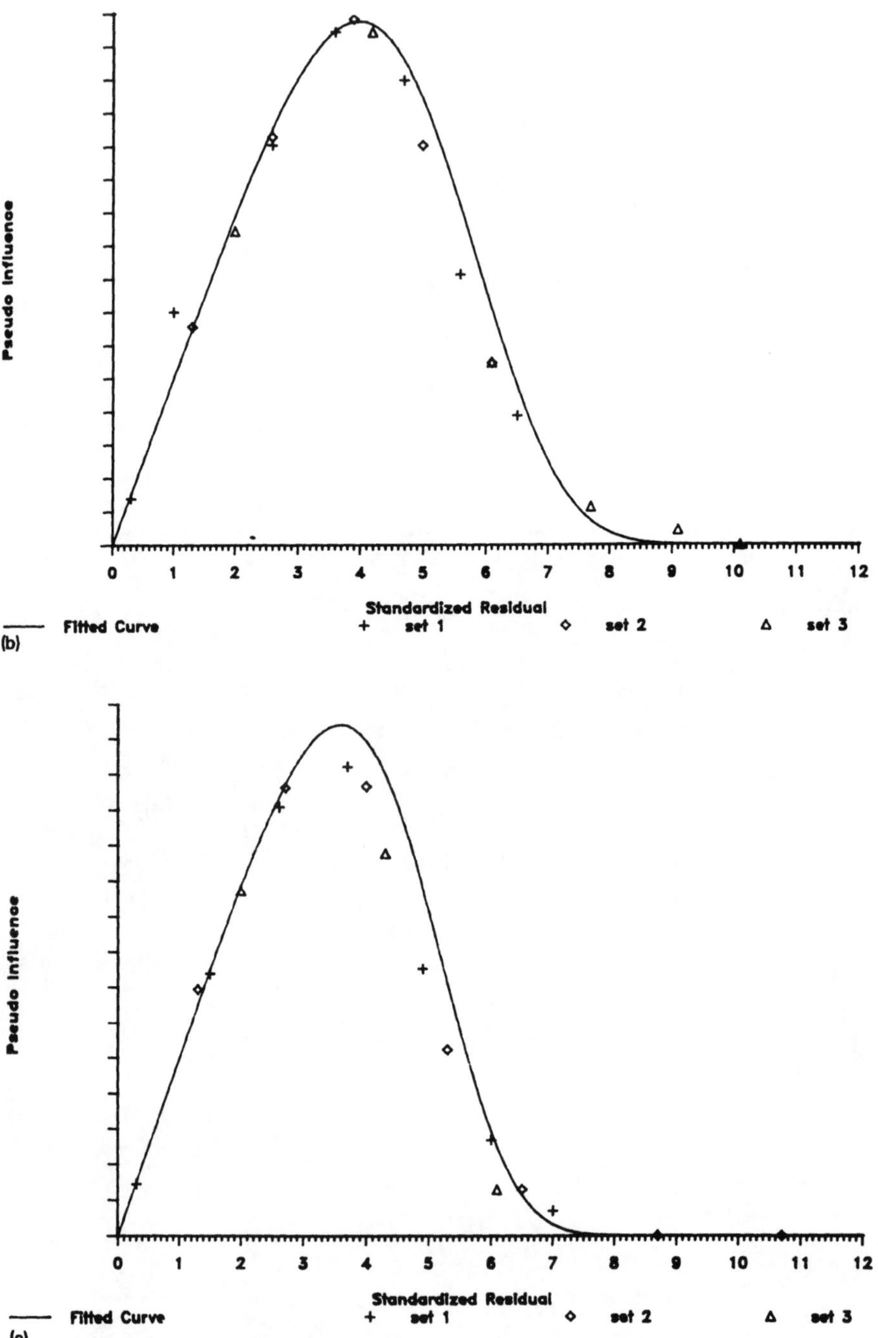

FIGURE 2 (Continued)

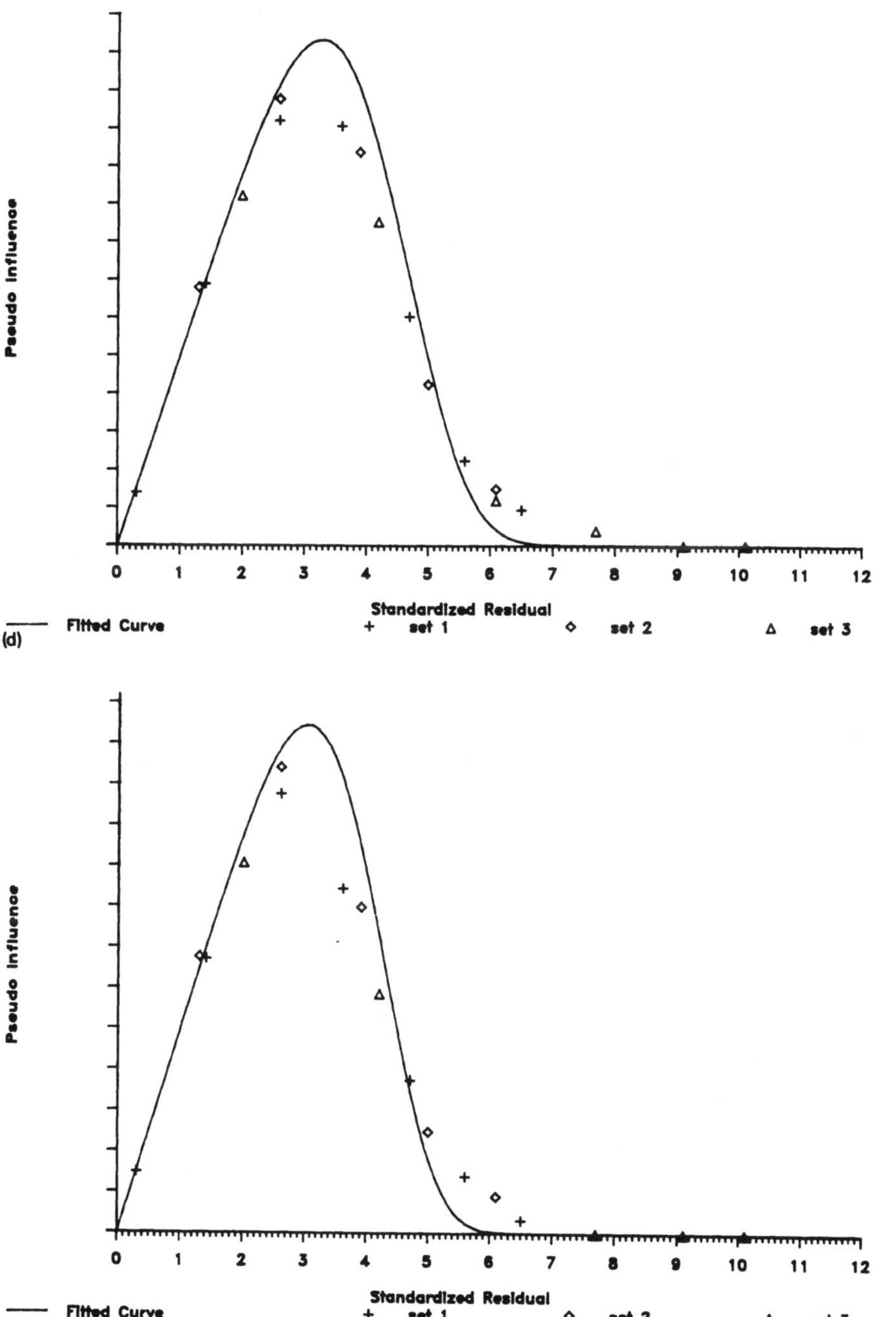

Pseudo Influence
0
1
2
3
4
5
6
7
8
9
10
11
12
Standardized Residual
Fitted Curve
set 1
set 2
set 3
(d)
Pseudo Influence
0
1
2
3
4
5
6
7
8
9
10
11
12
Standardized Residual
Fitted Curve
set 1
set 2
set 3
(e)

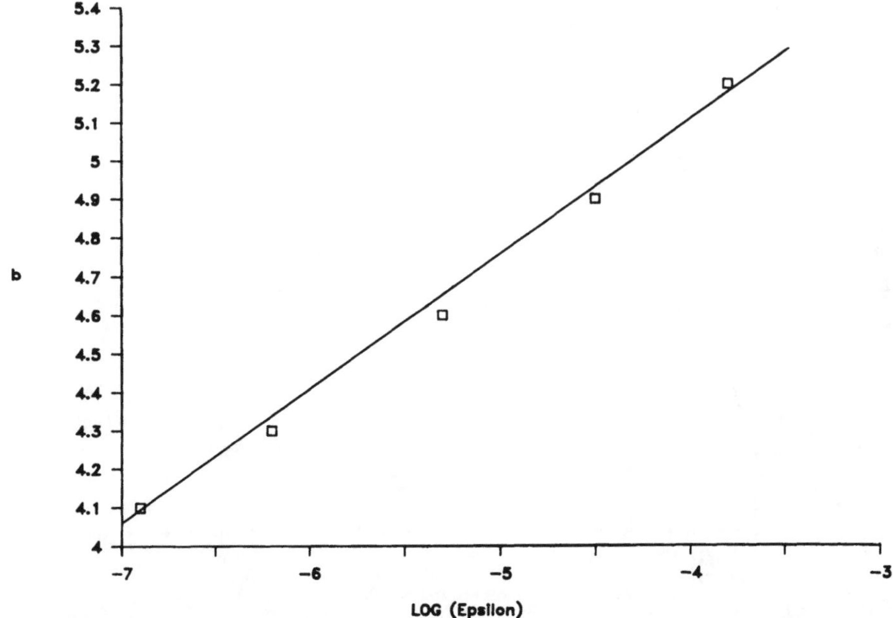

FIGURE 3 Relationship between the parameter b in the approximation function and the parameter ϵ(α and k) in the Bayesian prior distribution.

5. CONCLUSION

Box (1979) has argued that when a standard procedure is inadequate, it is the model, not the estimation procedure, that requires modification. In dealing with the "bad value" problem, a direct Bayesian analysis with a single contaminated model leads to weighting patterns very similar to those of M-estimators which are claimed by their proponents to work well in practice. This verifies the suggestion that if an appropriate model is employed, appropriate estimates will be obtained using standard estimation methods.

ACKNOWLEDGMENT

Sponsored by the United States Army under Contract No. DAAG29-80-C-0041.

REFERENCES

Andrews, D. F., Bickel, P. J., Hampel, F. R., Huber, P. J., Rogers, W. H., and Tukey, J. W. (1972). Robust Estimates of Location: Survey and Advances. Princeton University Press.

Barnett, V. D., and Lewis, T. (1978). Outliers in Statistical Data. Wiley, Chichester.

Beaton, A. F., and Tukey, J. W. (1974). The fitting of power series, meaning polynomials, illustrated on band-spectroscopic data Technometrics 16: 147-185.

Box, G. E. P. (1979). Some problems of statistics and everyday life J. Am. Stat. Assoc. 74: 1-4.

Box, G. E. P. (1980). Sampling and Bayes' inference in scientific modelling and robustness. J. R. Stat. Soc. Ser. A 143(4): 383-430.

Box, G. E. P., and Tiao, G. C. (1968). A Bayesian approach to some outlier problems. Biometrika 55(1): 119-129.

Chen, G. G., and Box, G. E. P. (1979). Implied assumptions for some proposed robust estimators. MRC Technical Summary Report 1997, University of Wisconsin, Madison.

Dixon, W. J. (1953). Processing data for outliers. Biometrics 9: 74-89.

Hogg, R. V. (1974). Adaptive robust procedures: A partial review and some suggestions for future applications and theory. J. Am. Stat. Assoc. 69: 909-922.

Huber, P. J. (1972). Robust statistics: A review. Ann. Math. Stat. 43: 1041-1067.

Jeffreys, H. (1932). An alternative to the rejection of observations. Proc. R. Soc. Ser. A 137: 78-87.

Tukey, J. W. (1960). A survey of sampling from contaminated distributions. Contributions to Probability and Statistics: Essays in Honor of Harold Hotelling. Standard University Press.

2

Some Properties of L_p-Estimators

VINCENT A. SPOSITO Iowa State University, Ames, Iowa

1. INTRODUCTION

Various generalized properties of L_p-estimators ($p \geqslant 1$) are underscored in this chapter. Efficient computational procedures are given which are based on some of these properties. Furthermore, some of these properties are shown to be useful in verifying if any Median-Polish fit is also an L_1 solution, i.e., which minimizes the sum of absolute deviations.

2. NOTATION AND PRELIMINARIES

Consider the model $y = X\beta + e$ where x_{ij}, $i = 1, 2, \ldots, N$, $j = 1, 2, \ldots, p$, denote a set of N observational measurements on p "independent variables," and y_i, $i = 1, 2, \ldots, N$, denote the associated measurements on the "dependent variable." Also, we shall let $x_{i1} = 1$ for $i = 1, 2, \ldots, N$. Using the criterion of minimizing the sum of absolute deviations or the L_1 criterion one desires to determine regression coefficients b_j which will

$$\text{minimize} \sum_{i=1}^{N} |y_i - \sum_{j=1}^{p} x_{ij}b_j|$$

For the linear model under consideration, let e_i^+ and e_i^- denote the vertical deviations "above" and "below" the fitted hyperplane for the ith observational set. By the nature of the problem, at most one deviation can be nonzero, and any fitted equation can be represented in terms of e_i^+ and e_i^-. In particular, for any i

$$\sum_j x_{ij}b_j + e_i^+ - e_i^- = y_i \tag{1}$$

By this construction $e_i^+ + e_i^-$ is the absolute deviation of the ith observation between the fit $\Sigma_j x_{ij}b_j$ and y_i.

In view of the above, one can easily formulate this problem as a linear programming problem where the objective function would be to minimize $\Sigma_{i=1}^N (e_i^+ + e_i^-)$. With (1) we have the following (primal) linear programming problem:

$$\begin{aligned} &\text{minimize} \quad 1'e^+ + 1'e^- \\ &\text{subject to} \quad Xb + Ie^+ - Ie^- = y \\ &\qquad\qquad\qquad e^+, e^- \geqslant 0 \end{aligned} \tag{2}$$

where 1 denotes the N-dimensional unit column vector. The dual problem of (2) is

$$\begin{aligned} &\text{maximize} \quad f'y \\ &\text{subject to} \quad X'f = 0 \text{ and } -1 \leqslant f_i \leqslant 1 \text{ for all } i \end{aligned} \tag{3}$$

An important relationship between the primal and dual problems is given by the theorem of complementary slackness (Arthanari and Dodge, 1981), which states that either an optimal nonnegative primal variable takes value zero or the corresponding dual constraint is satisfied as an equality (i.e., the corresponding dual slack variable takes value zero). For our problem this implies that

(i) either $e_i^+ = 0$ or $f_i = 1$

and

(ii) either $e_i^- = 0$ or $f_i = -1$

Hence, if the ith point is above a certain hyperplane, then $e_i^+ > 0$, so that $f_i = 1$; if this point is below this hyperplane, $e_i^- > 0$, so that $f_i = -1$.

The theorem of complementary slackness, as well as some fundamental results of linear programming, will be used in the next sections to establish some generalized properties under L_1.

3. GENERALIZED PROPERTIES OF L_1 ESTIMATORS

Property 1. If X is an N×p design matrix of rank k ($\leq$ p), then there exists some optimal hyperplane under L_1 which passes through at least k observation points.

Proof: Given by Barrodale and Roberts (1970), Theorem 1.

Property 2. Assume the maximum number of observations that lie on any optimal L_1 hyperplane is n^*. Then any optimal solution under L_1 is such that the absolute difference between the numbers of observations above and below this optimal hyperplane never exceeds n^*.

Proof: Let N^+ and N^- denote the numbers of observations above and below any optimal L_1 solution. Then from the theorem of complementary slackness:

(i) either $e_i^+ = 0$ or $f_i = 1$

and

(ii) either $e_i^- = 0$ or $f_i = -1$

Now, from the first constraint of the dual problem (3), we have that

$$\sum_{i=1}^{N} f_i = \sum_{i\varepsilon I^+} f_i + \sum_{i\varepsilon I^-} f_i + \sum_{i\varepsilon I^0} f_i = 0$$

where I^+, I^-, and I^0 denote the respective indices that lie above, below or on this optimal hyperplane. Therefore in view of (i) and (ii) with $f_i \varepsilon [-1, 1]$

$$|N^+ - N^-| + |\sum_{i\varepsilon I^0} f_i| \leq n^*$$

This bound can be obtained for certain data sets; in particular consider

x	1	2	3	3	4	5
y	0	1	0	1	1	0

then $\tilde{y} = 0$ and $\tilde{y} = 1$ are optimal L_1 solutions and $|N^+ - N^-| = n^* = 3$.

The following underscores several additional properties exhibited by L_1 estimators as noted by Gentle et al. (1977). Section 3 addresses two computational procedures for the simple regression model based on these properties.

In addition to Properties 1 and 2 the following properties hold in the general setting (see Gentle et al., 1977):

Property 3. No observation point lines between any two L_1 hyperplanes.

An L_1 hyperplane which passes through a maximal number of coplanar observations (at least k points), we shall denote as an extreme L_1 hyperplane. These hyperplanes are always generated using linear programming computational procedures at each iteration. Assume $n^* = p$ in Property 2; hence, we have the following properties:

Property 4. In the simple linear model, there exist at most two extreme lines that pass through a common observation point.

Property 5. If two extreme L_1 hyperplanes (lines) exist, then any convex combination is also optimal under L_1.

Consider now the situation where the sample size, N, is odd.

Property 6. All optimal hyperplanes under L_1 pass through a common observation point.

Property 7. There exist at most two extreme L_1 planes when N is odd.

The above two properties, in addition to Property 5, imply that any computational procedure needs only to identify at most two extreme lines in the simple model if one desires to generate the entire L_1 solution space when the sample size is odd.

Unfortunately, the situation when N is even does not follow in the same manner.

Property 8. If N is even, then there exist at most four extreme L_1 lines.

Property 9. If N is even, then there exist at most two extreme lines which pass through any given observation point.

Consider, for example,

x	1	1	2	2
y	1	2	1	2

then Properties 8 and 9 clearly are satisfied.

4. COMPUTATIONAL PROCEDURES

The properties given in the previous section imply that the search for an L_1 line for the simple linear model is restricted to a subset of $\binom{N}{k}$ lines where k denotes the rank of X. In addition, only lines such that $|N^+ - N^-| \leqslant n^*$ need be considered as possible extreme L_1 lines. An optimal L_1 line can easily be identified in view of the following results based on the theory of linear duality (Sposito, 1975):

Property 10. Any feasible extreme point solution of the dual problem (3) associated with a feasible solution of the primal L_1 problem (2) is an optimal solution under L_1.

Proof: Let f be any feasible extreme point solution of problem (3) associated with a feasible solution $(\beta^*, e^{*+}, e^{*-})$ of problem (2). Then

$$X\beta^* + Ie^{*+} - Ie^{*-} = y$$

so

$$f^{*\prime}(X\beta^* + Ie^{*+} - Ie^{*-}) = f^{*\prime}y \tag{4}$$

Also, f^* is such that $X'f^* = 0$; therefore (4) reduces to

$$f^{*\prime}(Ie^{*+} - Ie^{*-}) = f^{*\prime}y \tag{5}$$

Now, N^+ (N^-) of the e_i^{*+} (e_i^{*-}) are strictly positive, and moreover the remaining e_i^{*+} (e_i^{*-}) are zero. Therefore, by appealing to the theorem of complementary slackness,

$$\text{if } e_i^{*+} > 0, \text{ then } e_i^{*-} = 0 \text{ and } f_i^* = 1$$

and

$$\text{if } e_i^{*-} > 0, \text{ then } e_i^{*+} = 0 \text{ and } f_i^* = -1$$

This implies that the left-hand side of (5) can be expressed as

$$(1'Ie^{*+} + 1'Ie^{*-}) = \Sigma(e_i^{*+} + e_i^{*-})$$

Moreover,

$$\Sigma(e_i^{*+} + e_i^{*-}) = f^{*\prime}y$$

i.e., the optima of problems (2) and (3) are equal; therefore by the duality theorem of linear programming (Arthanari and Dodge, 1981), or Sposito, 1975), f^* and β^* are necessarily optimal solutions.

Consider, for example, the following set of observations:

x	0	1	1	2	2
y	2	3	1	3	1

Then, in view of the necessary properties exhibited by any extreme L_1 line, we can restrict our search to four linear equations, since $k = 2$ and $n^* = 2$.

Selecting $y = 2 + 1/2x$ we have

$$e_2^+ > 0 \text{ so that } f_2 = 1$$

$$e_3^- > 0 \text{ so that } f_3 = -1$$

and

$$e_5^- > 0 \text{ so that } f_5 = -1$$

Therefore, the constraints of the dual problem reduce to determining f_1 and f_4 such that

$$f_1 + f_4 = 1, \quad 2f_4 = 2, \quad f_1, f_4 \in [-1, 1]$$

Clearly $f_1 = 0$ and $f_4 = 1$ satisfy this system of constraints, hence $y = 2 + 1/2x$ is an extreme L_1 line.

Moreover, in view of Property 7 there exists at most one other L_1 extreme line solution. Considering in a like manner $y = 2 - 1/2x$ we see that this line is also optimal, and hence, by Property 5, any convex combination of these two lines is also optimal under L_1.

The following concluding properties are a direct consequence of the simplex procedure of linear programming (Sposito, 1975).

Property 11. If any of the derived dual variables are equal to ± 1, and these values are unique, then necessarily multiple solutions exist.

Property 12. If any of these derived (unique) variables are such that $|f_j| < 1$, then the L_1 estimator is unique.

The properties exhibited by L_1 estimators given above differ from the usual properties considered in a statistical setting.

However, it should be noted that in view of the primal linear programming problem (2), if the L_1 estimator $\tilde{\beta}$ is such that $X'(I\tilde{e}^+ - I\tilde{e}^-) = 0$, then $X'X\tilde{\beta} = X'y$; hence the L_1 estimator is necessarily an L_2 estimator.

Computational procedures to obtain L_1 estimators for the general linear model can be obtained by using Barrodale and Roberts's (1973) improved algorithm, or by a modification of this linear programming algorithm incorporating an L_2 estimator as discussed by McCormick and Sposito (1976).

Consider again the problem of determining estimates of the parameters of the simple linear model, $y = \alpha + \beta x + e$, under the criterion of L_1. This principle can be expressed as determining α and β which minimize

$$F = \sum_{i=1}^{N} |y_i - \alpha - \beta x_i| \tag{6}$$

As shown previously, estimates of α and β can be obtained under L_1 by reformulating (6) as a linear programming problem, (2) or (3). Armstrong and Kung (1978) showed that if one considers the dual problem then one can take advantage of the simple upper bound structure. We shall show that a faster approach can be taken which does not require linear programming techniques.

Weslowsky (1981) derived a computational procedure for obtaining estimates of (6) using a descent technique, hence solving (6) directly rather than formulating (6) as a linear programming problem or constructing its dual problem. We shall consider certain modifications of Wesolowsky's procedure which will reduce the overall computational effort.

Wesolowsky noted that (6) is a convex function composed of linear segments; hence the minimum will necessarily occur at the intersection of two lines. Given an initial point on any line, one can determine a "better" point, which will decrease the value of F by descending to a nearby intersection. Repeating this sequence, the procedure continues until a point is reached which minimizes (6). The motivation for Wesolowsky's procedure stems from the fact that an optimal L_1 plane will pass through two observation points, i.e., property 1 where rank X = 2. Hence, considering any point (x_j, y_j), minimizing $F = \Sigma_{i=1}^{N} |y_i - \alpha - \beta x_i|$ can be written equivalently as

$$\text{minimize} \sum_{i \neq j} |x_i - x_j| \, \left| \frac{y_i - y_j}{x_i - x_j} - \beta \right| \tag{7}$$

for a particular (x_j, y_j). Note if $x_i = x_j$ for any i, then necessarily the corresponding weight is defined to be zero.

Moreover, (7) can be expressed as

$$\text{minimize} \sum_{i=1}^{N} w_i |a_i - \beta| \tag{8}$$

where w_i are nonnegative constants and a_i are constants. Hence, (8) is equivalent to the problem of finding the weighted median of the a_i values using w_i as weights. The optimal weighted median for a given set of a_i's and w_i's is such that for an appropriate j:

$$\sum_{i=1}^{j} w_{(i)} \geq \sum_{j+1}^{N} w_{(i)}$$

and

$$\sum_{i=1}^{j-1} w_{(i)} < \sum_{i=j}^{N} w_{(i)}$$

where $w_{(i)}$ denotes the corresponding weights of the ordered a_i's.

Another modification of Wesolowsky's procedure considers the initial estimate of β; clearly any initial estimate would suffice, with a near-optimal L_1 estimate being more appropriate. The near-optimal estimates used in our study include $\tilde{\alpha} = \tilde{y}$ with $\tilde{\beta} = 0$, and the classical (closed form) L_2 solution.

In 1978, Armstrong and Kung showed that SUBROUTINE SIMLP (simple upper bounding procedure) was considerably more efficient in terms of total computational time than those of other known computational procedures in determining L_1 estimates of a simple linear model.

A Monte Carlo study was conducted to compare SUBROUTINE SIMLP with the above modifications of Wesolowsky's procedure for the two-parameter linear models. One thousand repetitions on random problems were generated as follows:

1. Generate at random a solution (α, β) where α and β are iid U(-4, 4).
2. Generate an X matrix where $x_{i1} = 1$ for all i and x_{i2} are iid U(-100, 100) for all i.
3. Generate the dependent variables such that $y_i = (\alpha + \beta x)_i + e_i$ where e_i are iid U(-1, 1) for all i.

Iowa State University's ITEL AS/6 was used in the study. Table 1 contains the summary of our study.

Subroutine DESL1 used $\tilde{\alpha} = \overline{y}$ with $\tilde{\beta} = 0$ as an initial estimate, whereas subroutine DESLS incorporated the closed-form least squares solution. For sample sizes less than 50, SUBROUTINE DESL1 is on the average slightly faster than SUBROUTINE SIMPLP; for larger sample sizes SUBROUTINE DESL1 is considerably faster. Moreover, using the L_2 estimator, SUBROUTINE DESLS is more efficient in reducing total CPU time than either of the other two subroutines.

A listing of SUBROUTINE DESL1 is given by Josvanger and Sposito (1983).

TABLE 1 Time Comparison Study

No. of observations	DESLS time (sec)	DESL1 time (sec)	SIMLP time (sec)	Avg. reduction over SIMLP	Best
10	12.40	12.39	12.41	.00002	DESL1
25	33.59	33.58	33.62	.00004	DESL1
30	41.61	41.52	43.19	.00167	DESL1
50	69.01	69.71	94.61	.02560	DESLS
250	437.71	446.27	648.39	.11068	DESLS

5. PROPERTIES OF L_p ESTIMATORS ($p \geq 1$)

Property 13. Problems of the form

$$\text{minimize} \sum_{i=1}^{N} |y_i - (X\beta)_i|^p$$

where $p \geq 1$ always have an optimal solution.

For example, consider the L_1 problem (with $p = 1$):

$$\text{minimize} \quad \Sigma\,(e_i^+ + e_i^-)$$

$$\text{such that} \quad X\beta + Ie^+ - Ie^- = y$$

$$e^+,\ e^- \geq 0$$

If $y_i \geq 0$ for all i, then necessarily $e^{*+} = y$, $\beta^* = 0$, $e^{*-} = 0$ denotes a feasible solution. Likewise, if $y_i \leq 0$ for all i, then multiplying all constraints by minus 1, we have that $e^{*-} = y$, $\beta^* = 0$, $e^{*+} = 0$ denotes a feasible solution. Moreover, since the objective function is bounded below by zero, the L_1 problem always has an optimal solution.

If we consider minimizing $\Sigma\,|y_i - (X\ \ )_i|^p$ for any finite $p \geq 1$, then we have the problem

$$\text{minimize} \quad \Sigma\,|e_i^+|^p + \Sigma\,|e_i^-|^p$$

$$\text{such that} \quad X\beta + Ie^+ - Ie^- = y$$

$$e^+,\ e^- \geq 0$$

which also always has a feasible solution, and where its objective function is bounded below by zero; hence the L_p ($p \geq 1$) problem always has an optimal solution.

The case when $p \longrightarrow \infty$ is commonly denoted as minimizing the maximum deviation or the L_∞ problem. This problem can be constructed as a linear programming problem by considering the problem:

$$\text{minimize} \quad e$$

$$\text{such that} \quad -e \leq (X\beta)_i - y_i \leq e \quad \text{for all } i$$

where e denotes the maximum deviation we wish to minimize. Noting initially that for any β^* there always exists an e^* which satisfies the interval restrictions, then the L_∞ problem always has a feasible solution. For example, $\beta^* = 0$ with $e^* = \max|y_i|$ is one such feasible solution. Moreover, the objective fjnction e is bounded below by zero; hence the L_∞ problem always has an optimal solution.

Let Z denote the number of zero residuals obtained employing the L_1 norm. Then we have the following property:

Property 14. Zero is the median of the residuals if and only if $|N^+ - N^-| \leqslant Z$.

Proof: Let $N^+(\xi) = \text{Card}\{i \mid d_i > \xi\}$, $N^-(\xi) = \text{Card}\{i \mid d_i < \xi\}$, and $Z(\xi) = \text{Card}\{i \mid d_i = \xi\}$, where Card denotes cardinality. Then, the median of the residuals is any value ξ such that

$$N^+(\xi) + Z(\xi) \geqslant N^-(\xi)$$

and

$$N^-(\xi) + Z(\xi) \geqslant N^+(\xi)$$

Hence, $|N^+ - N^-| \leqslant Z$.

The converse follows immediately.

Property 15. Let (β_0^*, β^*) be an optimal solution to a normal approximation problem for $p \geqslant 1$, and let $e^* = y - \beta_0^* \underset{\sim}{1} - X\beta^*$. Then zero is the norm statistic for the same value of p when one considers the respective associated residuals.

Proof: Consider any $p \geqslant 1$, where v^0 is some value which minimizes $N_p(e^* - v\underset{\sim}{1})$.

Therefore,

$$\begin{aligned} N_p(e^* - 0.1) &= N_p(e^*) \\ &= N_p(y - \beta_0^* \underset{\sim}{1} - X\beta^*) \\ &\leqslant N_p(y - [\beta_0^* + v]\underset{\sim}{1} - X\beta^*) && \text{for all v, since } (\beta_0^*, \beta^*) \text{ minimizes the respective norm problem} \\ &= N_p(y - \beta_0^* \underset{\sim}{1} - X\beta^* - v\underset{\sim}{1}) && \text{for all v} \\ &= N_p(e^* - v\underset{\sim}{1}) && \text{for all v} \end{aligned}$$

Hence, 0 minimizes $N_p(e^* - v\underset{\sim}{1})$.

Examples of Property 15:

1. There exists a median of the residuals of an optimal L_1 estimator which equals zero.
2. The mean of the residuals of the optimal least squares estimator is zero.
3. The midrange of the residuals with respect to the optimal L_∞ estimator is zero.

6. UNBIASED L_p REGRESSION ESTIMATORS

Consider the linear model $y = X\beta + e$ where x_{ij}, $1 = 1, 2, \ldots, N$, $j = 1, 2, \ldots, p$, denotes a set of N observational measurements on p independent variables, and y_i, $i = 2, 3, \ldots, N$, denotes the associated measurement on the dependent variable.

When one considers the estimation of β it is well known that the principle of least squares will yield an unbiased estimator when the vector of disturbances is symmetrically distributed. Moreover, this estimator is the "best linear unbiased estimator" provided the elements of e are uncorrelated and have the same variance.

Under L_1 (minimization of the sum of absolute residuals) and L_∞ (minimization of the maximum absolute deviation) the estimators obtained are nonlinear and moreover are usually nonunique even for linear models that are of full column rank. However, even if multiple solutions exist, it can be shown that a computational procedure will necessarily generate unbiased L_1 or L_∞ estimators if e is symmetrically distributed. This procedure is based on a condensed version of Sielken and Hartley's (1973) algorithm that requires the use of an antisymmetrical estimator $\tilde{\beta}_0$, that is, an estimator such that

$$\beta - \tilde{\beta}_0(e) = -\beta + \tilde{\beta}_0(-e)$$

Harvey (1978), also utilizing the notion of an antisymmetrical estimator, provides a simple proof of unbiasedness of the L_p estimator, $1 < p < \infty$, under the assumption that the linear model is of full column rank. In this situation uniqueness is assured. Hence, the L_p estimator will be unbiased regardless of the computational procedure used to produce it. We shall show that the unbiasedness of the L_p estimator for $p \geqslant 1$ can be obtained for the general situation where there may be nonunique estimators by appealing to the basic scheme of Sielken and Hartley. In this vein, it is assumed that the error distribution is symmetric with mean zero and the linear model is of arbitrary rank.

The L_p estimates for $p \geq 1$ can be obtained by finding a $\bar{\beta}$ that minimizes

$$\sum_{i=1}^{N} |y_i - (X\beta)_i|^p \tag{9}$$

Barrodale and Roberts (1970) noted that (9) can be converted into a convex programming problem by letting u_i denote the positive vertical deviations and v_i denote the negative vertical deviations, $i = 1, 2, \ldots, N$, about any fitted hyperplane. Hence, since at most one deviation can be nonzero for any ith observation, (9) can be written as

$$\sum_{i=1}^{N} (u_i^p + v_i^p) \text{ subject to}$$

$$X\beta + Iu - Iv = y, \qquad u, v \geq 0$$

Let us consider any antisymmetrical estimator $\tilde{\beta}_0$ and the following two equivalent convex programming problems:

P_1: minimize $\Sigma_{i=1}^{N} (u_i^p + v_i^p)$ subject to

$$[X, -X, I, -I] \begin{bmatrix} B_1 \\ B_2 \\ u \\ v \end{bmatrix} = [y - X\tilde{\beta}_0], \qquad B_1, B_2, u, v \geq 0$$

P_2: minimize $\Sigma_{i=1}^{N} (u_i^p + v_i^p)$ subject to

$$[X, -X, I, -I] \begin{bmatrix} B_2 \\ B_1 \\ v \\ u \end{bmatrix} = -[y - X\tilde{\beta}_0], \qquad B_1, B_2, u, v \geq 0$$

Unbiasedness can now be obtained by employing the technique outlined by Sielken and Hartley (1973):

1. Select one of the problems P_1 and P_2 with respective probabilities 1/2.
2. Determine an optimal solution of the problem selected obtaining $\bar{B}_1$ and $\bar{B}_2$.
3. Estimate β by $\bar{\beta} = \bar{B}_1 - \bar{B}_1 + \tilde{\beta}_0$. Since $y(-e) - X\tilde{\beta}_0(-e) = -y(e) + X\tilde{\beta}_0(e)$, then $P_1(-e)$ and $P_2(e)$ are identical except for the ordering of the variables; likewise for problems $P_1(e)$ and $P_2(-e)$. Therefore, letting $\bar{\beta} = \bar{B}_1 - \bar{B}_2 + \tilde{\beta}_0$ denote an optimal solution obtained by any algorithmic procedure, we have that

$$\bar{B}_1[P_1(-e)] - \bar{B}_2[P_1(-e)] = -\{\bar{B}_1[P_2(e)] - \bar{B}_2[P_2(e)]\}$$

and

$$\bar{B}_1[P_1(e)] - \bar{B}_2[P_1(e)] = -\{\bar{B}_1[P_2(-e)] - \bar{B}_2[P_2(-e)]\}$$

Furthermore, $E[\bar{\beta}] = \beta$, since

$$\begin{aligned} E[\beta] &= E[\bar{\beta} + \tilde{\beta}_0] \\ &= E[\bar{\beta} \mid e) + (\bar{\beta} \mid -e)] + \beta \\ &= E[(\bar{\beta} \mid e, P_1)1/2 + (\bar{\beta} \mid e, P_2)1/2 \\ &\quad + (\bar{\beta} \mid -e, P_1)1/2 + (\bar{\beta} \mid -e, P_2)1/2] + \beta \\ &= 0 + \beta \end{aligned}$$

Problems P_1 and P_2 are standard convex programming problems that that can be solved with any convex simplex procedure, for example, by Newton's method as discussed by Barrodale and Roberts (1970).

The above formulations, incorporating the least squares estimator, as in P_1 and P_2, have been used previously to reduce the total overall computational effort in Barrodale and Roberts's L_1 algorithm (1973) (see McCormick and Sposito, 1976, or Sposito et al., 1977).

7. OPTIMAL L_p ESTIMATORS BASED ON THE SAMPLE KURTOSIS

A frequent problem in statistics is that of estimating the mean of a symmetric distribution. For the model

$$y = \beta + e$$

one typically estimates β so as to minimize some vector norm of the residual vector $(y - \tilde{\beta})$. In particular, given a sample of size N, $\tilde{\beta}$ is determined so as to

$$\underset{\beta}{\text{minimize}} \sum_{i=1}^{N} |y_i - \beta|^p$$

this estimator is known as the L_p estimator. Rice and White (1964), as well as Ekblom and Henriksson (1969), emphasize that the properties of L_p estimators depend critically on the distribution of the residual vector. Rider (1957) considers several symmetric distributions and recommends that the sample midrange (L_∞ or Chebyshev estimator) is a better estimator of central tendency than $\bar{x}$, if the coefficient of kurtosis, k, is less than 2.2. Harter (1974–1976) proposed that if:

1. $k > 3.8$ use the median or L_1 estimator (minimize the sum of absolute deviations).
2. $k \in [2.2, 3.8]$ use the least squares estimator, the mean.
3. $k < 2.2$ use the Chebyshev estimator, the midrange.

Let us initially investigate how well the sample kurtosis approximates the population kurtosis and then recommend guidelines for selecting an optimal value of p based on the sample kurtosis.

Six symmetric distributions were considered to cover a fairly wide range of kurtosis. These distributions are:

1. Uniform: (kurtosis = 1.8)

$$f(x) = 1 \quad -1/2 \leq x \leq 1/2$$
$$= 0 \quad \text{otherwise}$$

2. Parabolic: (kurtosis = 2.14)

$$f(x) = .75(1 - x)^2 \quad -1 \leq x \leq 1$$
$$= 0 \quad \text{otherwise}$$

3. Triangle: (kurtosis = 2.40)

$$f(x) = 1 - |x| \quad -1 \leq x \leq 1$$
$$= 0 \quad \text{otherwise}$$

4. Normal: (kurtosis = 3.0)

$$f(x) = \frac{1}{\sqrt{2\pi\sigma_1^2}} e^{-1/2\left(\frac{x}{\sigma_1}\right)^2} \quad \text{for } -\infty < x < \infty \text{ where } \sigma_1^2 = 9$$

5. Contaminated normal: (kurtosis = 4.0 and 5.0)

$$f(x) = \frac{1}{\sqrt{2\pi\sigma_1^2}} e^{-1/2\left(\frac{x}{\sigma_1}\right)^2}$$

$$+ 1/2 \frac{1}{\sqrt{2\pi\sigma_2^2}} e^{-1/2\left(\frac{x}{\sigma_2}\right)^2} \quad \text{for } -\infty < x < \infty$$

This family of contaminated normal distributions was used to obtain error distributions with a specific kurtosis by choosing σ_1^2 and σ_2^2 so that the overall variance was 9. In particular, σ_1^2 and σ_2^2 were chosen to satisfy the system:

$$\sigma_1^4 + \sigma_2^4 = 54k$$

$$\sigma_1^2 + \sigma_2^2 = 18$$

For further details see Johnson and Kotz (1970).

6. Double exponential: (kurtosis = 6.0)

$$f(x) = \frac{1}{2} e^{-|x|} \quad \text{for } -\infty < x < \infty$$

To investigate whether the sample kurtosis

$$k = \frac{n\Sigma(y_i - \bar{y})^4}{\{\Sigma(y_i - \bar{y}^2\}^2}$$

provides a reasonable approximation of the population kurtosis, various sample sizes were selected at random from each of the above distributions. For each combination the procedure was repeated 500 times and the corresponding empirical average value is recorded in Table 2.

TABLE 2 Sample Kurtosis Based on Different Sample Sizes

		N				
Densities	k	30	50	100	200	400
Uniform	1.8	1.90	1.86	1.82	1.80	1.80
Parabolic	2.14	2.20	2.19	2.18	2.16	2.15
Triangle	2.40	2.44	2.42	2.41	2.40	2.40
Normal	3.0	2.93	2.98	2.99	3.00	3.00
Mixture of normals	4.0	3.56	3.61	3.74	3.82	3.99
	5.0	4.26	4.47	4.63	4.84	5.01
Laplace	6.0	4.47	4.99	5.27	5.64	6.01

A sample size of 400 was needed to obtain a value within .01 of the true population kurtosis. For finite populations a smaller sample size would suffice; for a sample size of 30, the computed empirical value is reasonable close to the population kurtosis for all the finite distributions considered. For long-tailed distributions a much larger sample size was needed, i.e., for the mixture of normals and the Laplace distribution.

To obtain an optimal value of p based on the population kurtosis (or sample kurtosis), a sample of size 400 was generated 100 times for each of the distributions considered. For each combination of error

TABLE 3 Empirical Mean Square Error of L_p-Norm Estimates (Sample Size of 400)[a]

	p				
k	1.00	1.25	1.50	1.75	2.00
3.0	.0432	.0379	.0345	.0257	<u>.0203</u>
4.0	.0309	.0255	<u>.0203</u>	.0218	.0224
5.0	<u>.0138</u>	<u>.0139</u>	.0158	.0178	.0218
6.0	<u>.0028</u>	.0033	.0036	.0041	.0048

[a] Source: From Sposito et al. (1983).

distribution and p-norm, the corresponding empirical mean square error was computed and is recorded in Table 3.

Based on these results, the following values of p for $p \in [1,2]$ is suggested:

$$p = 6/k$$

For $N \geqslant 200$ the above expression yields a value for p which is reasonably close to the optimal value of $p \in [1,2]$.

Money et al. (1982) suggested that

$$p = 9/k^2 + 1$$

represents an optimal value of p. Based on small sample sizes this expression yields a reasonable value of p for distributions with a finite range. For long-tailed distributions like the Laplace distribution this expression does not yield a value of p close to the optimal value of 1; hence a large sample is needed.

For small sample sizes and $p \in [1,\infty)$ a slight modification of Harter's results is suggested:

1. Use the Chebyshev estimator (L_∞ estimator) if $k < 2.2$.
2. Use the least squares estimator if $k \in [2.2, 3]$.
3. If $k \in (3, 6)$, use $p = 1.5$.

8. MEDIAN-POLISH AND L_1 ESTIMATORS

Suppose the response variables, y_{ij}, are affected by t levels of one factor and s levels of another factor. If one assumes that the responses can be expressed as the sum of separate contributions of the factors, then we have the following additive model:

$$y_{ij} = \mu + \alpha_i + \beta_j + e_{ij} \qquad i = 1, 2, \ldots, t, \quad j = 1, 2, \ldots, s$$

As is well known, least squares estimation is optimal when the e_{ij}'s are iid from a Gaussian distribution having mean zero and common variance.

Tukey (1977) and Hoaglin et al. (1983) noted that if there exist outliers or there exist a few "bad" data values in the data, then using a procedure called Median-Polish would yield better estimates than using the mean in two-way tables.

This procedure operates iteratively by subtracting the median of each row from each observation in that row and then subtracting the median from each column from this updated table. This sequence is continued until the median of each row and column is zero. This computational procedure, as noted by Tukey, could be started just as well with columns.

Median-Polish is a resistant technique; hence isolated values (large or small) in a small number of cells will not affect the estimates of μ, α_i, and β_j. As noted by Tukey, in the situation where the median in any row or column is not unique, then one could use the low-median.

To illustrate this technique consider the following two-way table.

$$\begin{bmatrix} 1 & 8 & 3 \\ 5 & 9 & 2 \\ 6 & 4 & 7 \end{bmatrix}$$

Starting with rows, we obtain

			Median
1	8	3	3
5	9	2	5
6	4	7	6

yielding

				Previous median
	-2	5	0	3
	0	4	-3	5
	0	-2	1	6
column median	0	4	0	5

Here, besides considering the median of each column, the median of the row effects is obtained, i.e., 5.

Subtracting the median of each column, including the bordering column, we now have the following table:

-2	1	0	-2
0	0	-3	0
0	-6	1	1
0	4	0	5

Since the median (of the residuals) in every row and column is zero, the procedure terminates with

1. The overall common value of 5, ($\tilde{\mu}$).
2. The row effects, $\tilde{\alpha}$: (-2, 0, 1).
3. The column effects, $\tilde{\beta}$: (0, 4, 0).

Moreover, the following entries of this final table:

$$\begin{bmatrix} -2 & 1 & 0 \\ 0 & 0 & -3 \\ 0 & -6 & 1 \end{bmatrix}$$

denote the residuals obtained using the Median-Polish fit procedure.

Several studies have underscored many computational aspects of Median-Polish (Bloomfield and Steiger, 1983); Siegel, 1983). Moreover, as noted by Tukey (1977), the final Median-Polish fit is often close to an L_1 solution or could terminate on an L_1 solution. This follows since min $\Sigma\Sigma|e_{ij}|$ under Median-Polish is bounded below by min $\Sigma\Sigma|e_{ij}|$ under L_1.

Siegel (1983) showed that for 3×3 tables, if the first half-step of Median-Polish results in a row or a column of zeros, then necessarily Median-Polish will converge in at most three steps to an L_1 solution. Bloomfield and Steiger (1983) extended Siegel's results for 3×3 tables by using a certain modified Median-Polish; in particular, in the first half-step the value subtracted from each row or column in the first entry in that row or column. This resulted in the following (Bloomfield and Steiger, 1983, p. 124):

> For any 3×3 table, modified Median-Polish will converge in at most three steps to an L_1 solution.

They further note that this procedure cannot be regarded as an algorithm to obtain L_1 solutions in larger tables.

Hoaglin et al. (1983, p. 186) note that several questions in the area of Median-Polish remain open; in particular, how can one recognize tables for which the final Median-Polish solution and an optimal L_1 solution differ? More specific to our discussion:

> How can one determine if any Median-Polish fit is equivalent to an L_1 solution?

We will address this question and show that there exists a simple sufficient condition to check this equivalence. This condition follows by appealing to results given by Sposito (1975, Chapter 5) which are based on the Kuhn-Tucker optimality conditions.

As noted by Barrodale and Roberts (1973) and others, the L_1 problem can be expressed as the following linear programming problem:

$$\begin{aligned} &\text{minimize} \quad \Sigma e_i^+ + \Sigma e_i^- \\ &\text{such that} \quad X\beta + Ie^+ - Ie^- = y \\ &\qquad\qquad\qquad e^+, e^- \geq 0 \end{aligned} \tag{10}$$

where e_i^+ (and e_i^-) denote the vertical deviations above (and below) the ith observation about any fitted equation.

An important problem associated with (10) is the following dual problem:

$$\begin{aligned} &\text{maximize} \quad y'f \\ &\text{such that} \quad X'f = 0 \\ &\qquad\qquad \left.\begin{aligned} f_i &\leq 1 \\ -f_i &\leq 1 \end{aligned}\right\} \text{for all } i \end{aligned} \tag{11}$$

The following relationships hold between (10) and (11), as shown by Sposito (1975).

Lemma 1. If $(\beta^*, \bar{e}^+, \bar{e}^-)$ is any feasible solution of (10), i.e., satisfies the set of restrictions in (10), and f^* is any feasible solution of (11) such that

$$\Sigma \bar{e}^+ + \Sigma \bar{e}^- = y'f^*$$

then necessarily these solutions are optimal solutions.

Lemma 2. For any optimal solution of (10), $(\beta^0, \bar{e}^+, \bar{e}^-)$, there exists an optimal solution of (11), f^0. Moreover,

$$\text{(a)} \quad \bar{e}_i^+ (f_i^0 - 1) = 0$$

$$\text{(b)} \quad \bar{e}_i^- (-f_i^0 - 1) = 0$$

Lemma 2 illustrates that if $e_i^+ > 0$ about $\tilde{\beta}$, then necessarily the corresponding dual variable $f_i = 1$ if β is indeed an L_1 solution; moreover, if $e_i^- > 0$, then $f_i = -1$. Note that if one was considering the true residuals, then $-e_i^- = r_i$, that is, denotes a negative residual.

Appealing to the above results, one is able to establish a sufficient condition to verify if any proposed solution (hyperplane) is an L_1 solution. This is given in the following property:

Property 16. If any proposed L_1 hyperplane is such that

$$f_i^* = 1 \quad \text{for } \bar{e}_i^+ > 0$$

$$f_i^* = -1 \quad \text{for } \bar{e}_i^- > 0$$

and the remaining f_i's satisfy the feasible region of problem (11), then this proposed hyperplane is an L_1 solution.

Proof: Let $f_i^* = 1$ for $\bar{e}_i^+ > 0$ and $f_i^* = -1$ for $\bar{e}_i^- > 0$ where $\bar{e}_i^+$ $(\bar{e}_i^-)$ are the vertical deviations about the proposed L_1 hyperplane. Moreover, assume the remaining f_i^*'s are such that $X'f^* = 0$ with $-1 \leq f_i^* \leq 1 \; \forall_i$

Now, the proposed L_1 solution $(\bar{\beta}, \bar{e}^+, \bar{e}^-)$ is such that

$$f^{*\prime}(X\beta + I\bar{e}^+ - I\bar{e}^-) = f^{*\prime}y \tag{12}$$

since f^* is a feasible solution of the dual problem; then necessarily $X'f^* = 0$, hence (12) can be rewritten as

$$f^{*\prime}(I\bar{e}_i^+ - I\bar{e}_i^-) = f^{*\prime}y \tag{13}$$

Now, N_1 of the e_i^+'s and N_2 of the e_i^-'s are positive with the remaining $\bar{e}_i^+ = 0 = \bar{e}_i^-$. Hence, it follows that

$$f^{*\prime}(I\bar{e}^+ - I\bar{e}^-) = 1'I\bar{e}^+ + 1'I\bar{e}^- = \Sigma(\bar{e}_i^+ + \bar{e}_i^-) \tag{14}$$

where 1 is a unit vector.

In particular, (13) and (14) imply that the optima of the L_1 problem and its dual problem are equal; hence, by Lemma 1, $(\beta^*, \bar{e}^+, \bar{e}^-)$ necessarily solves the L_1 problem.

To illustrate how one can use these results to determine whether a final fit under Median-Polish is also an L_1 solution, assume the following residuals were obtained using Median-Polish:

$$\begin{bmatrix} -2 & 0 & 0 \\ 0 & 6 & -1 \\ 7 & 0 & 0 \end{bmatrix}$$

To check if this Median-Polish is also an L_1 solution, one need only follow the steps listed below.

Since this final table denotes the true residuals, r_{ij}, obtained with Median-Polish, we have that r_{11} $(= e_1^-) = -2 < 0$, therefore $f_{11} = -1$. Likewise, since $r_{22} = 6$, then necessarily $f_{22} = 1$; also, $r_{23} = -1$ implies that $f_{23} = -1$ and $r_{31} = 7$ implies that $f_{31} = 1$.

It now remains only to verify that the remaining f_{ij}'s satisfy the dual constraints to verify that we have also an L_1 solution, in particular

$$X'f = \begin{bmatrix} 1 & 1 & 1 & 0 & 0 & 0 & 0 & 0 & 0 \\ 0 & 0 & 0 & 1 & 1 & 1 & 0 & 0 & 0 \\ 0 & 0 & 0 & 0 & 0 & 0 & 1 & 1 & 1 \\ 1 & 0 & 0 & 1 & 0 & 0 & 1 & 0 & 0 \\ 0 & 1 & 0 & 0 & 1 & 0 & 0 & 1 & 0 \\ 0 & 0 & 1 & 0 & 0 & 1 & 0 & 0 & 1 \end{bmatrix} f = 0 \tag{16}$$

with $f_{ij} \varepsilon [-1,1]$ for i, j and with $\Sigma_i \Sigma_j f_{ij} = 0$. This implies that the remaining set of $\{f_{ij}\}$ must be such that

$$f_{12} + f_{13} = 1$$

$$f_{21} = 0$$

$$f_{32} + f_{33} = -1$$

$$f_{21} = 0$$

$$f_{12} + f_{32} = -1$$

$$f_{13} + f_{33} = 1$$

with $-1 \leq f_{ij} \leq 1$ for these remaining i, j with $\Sigma_i \Sigma_j f_{ij} = 0$. Since $f_{12} = 0$, $f_{13} = 1$, $f_{32} = -1$ with $f_{33} = 0$, all these derived variables are between -1 and 1 and moreover $\Sigma_i \Sigma_j f_{ij} = 0$. Therefore, necessarily the obtained Median-Polish fit is also an L_1 solution.

In the situation where one follows the above simple steps and the derived f_{ij}'s are not all between -1 and 1, then necessarily the Median-Polish fit is not an L_1 solution.

It should be noted that the above sufficiency conditions can also be verified by considering (15) rather than (16), since there exists a one-to-one correspondence between the two, as can be seen by inspection of X'. In particular, the residuals in (16), i.e.,

$$\begin{bmatrix} -2 & 0 & 0 \\ 0 & 6 & -1 \\ 7 & 0 & 0 \end{bmatrix}$$

give us that the known components of the $\{f_{ij}\}$ matrix initially must be such that

$$\begin{bmatrix} -1 & & \\ & 1 & -1 \\ 1 & & \end{bmatrix}$$

Furthermore, the remaining of f_{ij}'s must be such that the sum of each column and row must equal zero $\forall\ f_{ij} \in [-1, 1]$, i.e.,

-1	f_{12}	f_{13}	= 0
f_{22}	1	-1	= 0
1	f_{32}	f_{33}	= 0
0	0	0	

By inspection we have that

$$\begin{bmatrix} -1 & 0 & 1 \\ 0 & 1 & -1 \\ 1 & -1 & 0 \end{bmatrix}$$

hence, the final Median-Polish fit given in (16) is also an L_1 solution.

The above sufficient condition can also be used for two-way tables with multiple cell entries. Consider the following simple table:

		Median
17, 12	11	12
7	8	

where the median of row one is unique and equals 12 and the low-median is selected in row 2. Subtracting these row median values, we have the following table:

5, 0	-1
0	1

(17)

Clearly, since not all the row and column medians are zero, technically we have not obtained a Median-Polish fit; however, the procedure illustrated can be used to verify whether or not one has a L_1 solution.

To establish this equivalence, one must determine if the remaining f_{ij}'s are such that in the following table all the row and column sums are equal to zero.

1 $f_{11}(2)$	-1
f_{22}	1

Since letting $f_{11(2)} = 0$ with $f_{22} = -1$, we have from Property 16 that necessarily this Median-Polish fit is also an L_1 solution. One should also note that on using either the low-median or the high-median in the next half-step of Median-Polish on (17), one can identify two other L_1 solutions.

9. ASYMPTOTIC DISTRIBUTION OF THE L_1 ESTIMATOR

Bassett and Koenker (1978) developed the asymptotic theory of the L_1 estimators and confirmed the conjecture of several studies that for any error distribution for which the L_1 estimator is "superior" to the mean as an estimator of location, the L_1 estimator is preferable. They showed that asymptotically

$$\sqrt{n}(\tilde{\beta} - \beta) \sim N(0, \lambda^2 n(X'X)^{-1}$$

where X is of dimension n×p, and where λ^2/n is the variance of the median from a sample of size n from a specified disturbance distribution.

A direct consequence of the results of Koenker and Bassett is that confidence intervals can easily be constructed. For example, a single component of β, β_i, has the following $(1 - \alpha)$ confidence interval:

$$\tilde{\beta}_i \pm z_{\alpha/2}\lambda(X'X)_{ii}^{-1/2}$$

where $(X'X)_{ii}^{-1/2}$ denotes the square root of the ith diagonal element of the matrix $(X'X)^{-1}$ and $z_{\alpha/2}$ denotes the appropriate percentile of the standard normal distribution.

Other inference procedures or hypothesis testing procedures can also be established in view of the Bassett and Koenker results, as discussed by Dielman and Pfaffenberger (1982). One major drawback, however, of using these results is that λ is unknown. Hence, an estimate of λ is needed to make the above approaches operational.

As shown by Cramer (1946, p. 369) asymptotically

$$\lambda = \{2f(m)\}^{-1} \tag{18}$$

where f(m) is the ordinate of the error distribution $\underset{\sim}{e}_t$ at the median, m. Denoting the L_1 estimator as $\tilde{\beta}$ so that $\tilde{e}_t = y_t - X_t\tilde{\beta}$, where X_t denotes the tth row vector of X, then $(f(m))^{-1}$ can be estimated by (Cox and Hinkley, 1974, p. 470):

$$(\tilde{f}(m))^{-1} = \frac{\tilde{e}_{(t)} - \tilde{e}_{(s)}}{(t - s)/n} \tag{19}$$

where $\tilde{e}_{(t)}$ denotes the ordered residuals of $\tilde{e}_t$. Moreover, they emphasize that t and s should be symmetric about the index of the median sample residual and the difference between t and s should be kept fairly small; in particular, $t = [n/2] + v$ and $s = [n/2] - v$ where $[\cdot]$ denotes the greatest integer value and where v is an appropriate positive integer. They also show that $(\tilde{f}(m))^{-1}$ is a consistent estimator of 2λ. No further comments were provided regarding λ; hence, we shall investigate how well (19) estimates $(f(m))^{-1}$ when f is a known symmetric distribution. A similar study was conducted by Dielman and Pfaffenberger (1982).

A Monte Carlo study was conducted to assess the usefulness of estimating λ in (18), appealing to (19).

Several symmetric distributions were considered:

1. Laplace

   $f(x) = 1/2 \exp\{-|x|\}$ for $-\infty < x < \infty$

2. Cauchy

   $f(x) = [\pi(1 + x^2)]^{-1}$ for $-\infty < x < \infty$

3. Normal

   $f(x) = (2\pi)^{-1/2} \exp\{-x^2/2\}$ for $-\infty < x < \infty$

4. Triangle

$$f(x) = \begin{cases} 1 - |x| & \text{for } -\infty \leqslant x \leqslant 1 \\ 0 & \text{otherwise} \end{cases}$$

5. Uniform

$$f(x) = \begin{cases} 1 & \text{for } -1/2 \leqslant x \leqslant 1/2 \\ 0 & \text{otherwise} \end{cases}$$

Without loss of generality, the model $y = \beta + e$ was considered since we are interested only in the term λ^2; in the general linear model $y = X\beta + e$ one is still interested only in λ^2. For the five symmetric distributions consider initially Table 4.

To determine the appropriateness of (19), random deviates from the above five distributions were generated with various sample sizes. $(f(0))^{-1}$ was then estimated for various values of $v = 1/2\,[\alpha n]$ for different $\alpha \in (0,1)$. Hence $t - s$ in (19) is equal to $[\alpha n]$.

As shown in Table 4, $[\alpha n]$ was set at 2, 4, 6, ..., 12. 300 repetitions were based on each specified combination and the results are summarized in Tables 5 and 6. Underlined entries in Tables 5 and 6 denote that the |estimate - true| $\leqslant$ .05.

(19) estimates the true value of $(f(0))^{-1}$ quite well for all the finite-range distributions considered. For these distributions, a value of $[\alpha n] = 6$ to 8 appears adequate in approximating $f(0))^{-1}$ for sample sizes of 100. For larger sample sizes ($\geqslant$200), $[\alpha n] = 2$ or 4 adequately approximates $(f(0))^{-1}$.

TABLE 4 Exact Values for Different Sample Sizes[a]

			$[4f^2(0)n]^{-1}$			
	$(f(0))^{-1}$	$[4f^2(0)]^{-1}$	n = 50	n = 100	n = 200	n = 300
Laplace	2.	1.	.02	.01	.005	.00332
Cauchy	3.1416	2.4674	.04935	.02467	.01234	.00822
Normal	2.5066	1.5708	.03142	.01571	.00785	.00524
Triangle	1.	.25	.005	.0025	.00125	.00083
Uniform	1.	.25	.005	.0025	.00125	.00083

[a] Source: From Sposito et al. (1983)

TABLE 5 Estimated Values of $(f(0))^{-1}$, Finite-Range Distributions[a]

		[αn]						
	$f(0))^{-1}$	2	4	6	8	10	12	n
Triangle	1.	1.053	1.060	1.062	1.063	1.069	1.079	50
		1.051	1.054	1.048	1.050	1.064	1.072	100
		1.045	1.029	1.045	1.059	1.071	1.079	200
		1.017	1.043	1.052	1.057	1.073	1.077	300
Uniform	1.	.991	.994	.990	.983	.981	.981	50
		1.009	1.006	.994	.988	.992	.989	100
		1.014	.993	1.000	1.004	1.006	1.002	200
		.989	1.009	1.010	1.005	1.009	1.009	300

[a] Source: From Sposito and Tveite (1986).

The estimation of $(f(0))^{-1}$ was very adequate in the case of the normal distribution, as seen in view of Table 6, for small values of $[\alpha n]$ over all four sample sizes considered. The estimation of $(f(0))^{-1}$ for the Laplace was adequate for sample sizes equal to 100 or larger and $[\alpha n] \geq 6$; for the Cauchy, however, a large value of n (= 300) was needed. For n = 50 and 100, the estimates of $(f(0))^{-1}$ generally increase for all values of $[\alpha n]$ for both Laplace and Cauchy distributions. For n = 200 or 300, the estimates of $(f(0))^{-1}$ decreased in value for the Laplace distribution and increased for the Cauchy distribution.

In summary, good estimates were obtained using (19) and $[4f^2(0)n]^{-1}$ for all the finite-range distributions considered as well as for the normal distributions for small value of $[\alpha n]$. For the Laplace and Cauchy distributions a larger sample size was needed to obtain reasonable estimates of $(f(0))^{-1}$ or $[4f^2(0)n]^{-1}$ for small values of $[\alpha n]$.

In general, a modest sample size is needed if one is using Cox and Hinkley's estimate of $(f(0))^{-1}$. This estimate can be used in the asymptotic distribution results of Bassett and Koenker (1978) provided the distribution can be assumed to be finite or nearly normal. In all the other cases, a very large sample should be used to provide reliable results in any L_1 inference situation.

The conclusions of the above study were based on the model y = β + e, hence yielding a single zero residual. In linear models with p

TABLE 6[a] Estimated Values of $(f(0))^{-1}$, Infinite-Range Distributions

		[αn]						
	$(f(0))^{-1}$	2	4	6	8	10	12	n
Laplace	2.	1.876	1.885	1.907	1.918	1.928	1.934	50
		1.939	1.934	1.960	1.956	1.970	1.980	100
		2.020	2.018	1.993	1.988	1.978	1.977	200
		1.994	1.980	1.977	1.974	1.963	1.958	300
Cauchy	3.1416	3.237	3.287	3.300	3.319	3.37	3.376	50
		3.196	3.258	3.294	3.317	3.333	3.402	100
		3.212	3.231	3.242	3.260	3.309	3.330	200
		3.098	3.181	3.181	3.220	3.261	3.292	300
Normal	2.5066	2.522	2.533	2.530	2.523	2.519	2.542	50
		2.546	2.546	2.520	2.513	2.532	2.539	100
		2.551	2.503	2.528	2.545	2.558	2.561	200
		2.488	2.541	2.549	2.544	2.565	2.571	300

[a] Source: From Sposito and Tveite (1986).

parameters, we conjecture that suitable values of v - s will be slightly larger than p + 1 since the number of zero residuals under L_1 (or LAV) is at least p, as discussed by Sposito et al. (1980).

10. EQUIVALENCE OF L_1 AND R-ESTIMATORS

Jaeckel (1972) showed that if the density function f(e) has finite Fisher-Information, the covariance matrix of X is positive definite and $(\underline{x}_i, y_i) \in R^{p+1}$ is an iid sample from the intercept model $y = X\beta + e$ with sign scores d_i satisfying

$$d_i = 0$$

where $d_1 \leq d_2 \leq \cdots \leq d_n$, then the R-estimator $\tilde{\beta}^*$, which minimizes

$$d_i(y - X\beta)_{(i)} \tag{20}$$

is such that

$$n^{1/2}(\tilde{\beta}^* - \beta) \xrightarrow{L} N(0, (X'X)^{-1}[2f(0)]^{-2}$$

Here,

$$d_i = \begin{cases} -1 & \text{if } i < (n+1)/2 \\ 0 & \text{if } i = (n+1)/2 \\ 1 & \text{if } i > (n+1)/2 \end{cases} \tag{21}$$

We shall denote the ith residual as $r_i = y_i - (X\tilde{\beta}^*)_i$.

<u>Property 17.</u> Let $(\underline{x}_i, y_i)$ R^{p+1}, $i = 1, 2, \ldots, n$, be a sample of size n from the intercept model $y = X\beta + e$ and assume $\underline{\alpha} \in R^p$ minimizes (20) where the sign scores d_i are given by (21). Let

$$\underline{\beta}(\underline{\alpha}) = (\alpha_1 + \text{med } r_j(\underline{\alpha}), \alpha_2, \alpha_3, \ldots, \alpha_p)$$

then $\underline{\beta}(\underline{\alpha}) = \tilde{\beta}$ is an L_1 estimator of β.

<u>Proof:</u>

$$\begin{aligned} \Sigma |r_i(\underline{\beta}(\underline{\alpha}))| &= \Sigma |r_i(\alpha_i + \text{med } r_j(\underline{\alpha}), \alpha_2, \ldots, \alpha_p)| \\ &= \Sigma |y_i - X(\alpha_1 + \text{med } r_j(\underline{\alpha})), \alpha_2, \ldots, \alpha_p)| \\ &= \Sigma |y_i - X\underline{\alpha} - \text{med } r_j(\underline{\alpha})| \\ &= \Sigma |r_i(\underline{\alpha}) - \text{med } r_j(\underline{\alpha})| \end{aligned}$$

therefore

$$\Sigma |r_i(\underline{\beta}(\underline{\alpha})| = \Sigma d_i(r_{(i)}(\underline{\alpha}) - \text{med } r_j(\underline{\alpha})$$

Now, since $r_j(\underline{\alpha}) = 0$ and since $\underline{\alpha}$ minimizes (20),

$$|r_i(\underline{\beta}(\underline{\alpha}))| = \Sigma\, d_i r_{(i)}(\underline{\alpha})$$
$$\leq \Sigma\, d_i r_{(i)}(\underline{\beta}) \quad \text{for any } \underline{\beta}$$
$$\leq \Sigma\, |r_{(i)}(\underline{\beta})| \quad \text{for any } \underline{\beta}$$

Therefore, among all $\underline{\beta}$, the $\underline{\beta}$ that minimizes the sum of absolute deviations is the $\underline{\beta}(\underline{\alpha})$ given above where α minimizes (20).

To illustrate the above results, consider the following location model $y = \beta_0 + e$ where for a sample of size three $y' = (1, 2, 5)$. Then minimizing $\Sigma_{i=1}^{3} d_i r_{(i)}(\alpha)$, we have that

d_i	$r_i(\alpha_1)$
-1	$1 - \alpha_1$
0	$2 - \alpha_1$
1	$5 - \alpha_1$

Hence, since $\Sigma_{i=1}^{3} d_i r_{(i)}(\alpha_1) = 4$, any value of α_1 is an R-estimator. Consider $\alpha_1 = 0$ and $\alpha_1 = 2$; then

$r_i(0)$	$r_i(2)$
1	-1
2	0
5	3

Both R-estimators yield an L_1 estimator; in particular $\widetilde{\beta} = \text{med } y_i = 2$. Moreover, for $\alpha_1 = 0$, $\underline{\beta}(\underline{\alpha}) = (\alpha_1 + \text{med } r_i(0)) = 0 + 2 = 2$ and for $\alpha_1 = 2$, $\underline{\beta}(\underline{\alpha}) = (\alpha_1 + \text{med } r_i(2)) = 2 + 0 = 2$.

In the same spirit, consider the simple regression model $y = \beta_0 + \beta_1 x + e$ where for a sample of size three:

x	y
1	1
2	2
3	5

The unique L_1 solution is $y = -1 + 2x$. Consider $\tilde{y} = (-1 - a) + 2x$; then

x	y	$r_i(a)$
1	1 - a	a
2	3 - a	-1 + a
3	5 - a	a $\longrightarrow$ med $r_i(a)$

Any value of a will yield an R-estimator since minimizing $d_i r_{(i)}(a) = -a + 0\ (-1 + a) = -1$. Consider

$$\alpha_1 = -1 - a$$

with

$$\alpha_2 = 2$$

then $\tilde{\alpha}_1 = \alpha_1 + \text{med } r_i(a) = (-1 - a) + a = -1$ and $\tilde{\alpha}_2 = \alpha_2 = 2$; that is, $\underline{\alpha}$ equals the L_1 estimator.

REFERENCES

Andrews, D. F., Bickel, P. J., et al. (1972). Robust Estimations of Location. University Press, Princeton, N. J.

Armstrong, R. D., and Kung, M. T. (1978). Least absolute value estimates for a simple regression problem. AS132, Appl. Stat. 27(3): 363-366.

Armstrong, R. D., and Frome, E. L. (1979). Least-absolute values estimation for one-way and two-way tables. Naval Res. Logistics Quarterly 26: 79-96.

Arthanari, T. S., and Dodge, Y. (1981). Mathematical Programming in Statistics. Wiley, New York.

Ashar, V. G., and Wallace, T. G. (1963). Sampling study of minimum absolute deviations estimators. Operations Res. 11: 747-758.

Banks, S. C., and Taylor, H. L. (1980). A modification to the discrete L_1 linear approximation algorithm of Barrodale and Roberts. SIAM J. Sci. Stat. Comput. 1(2): 187-190.

Barrodale, I., and Roberts, F. D. K. (1970). Applications of mathematical programming to ℓ_p approximation. Nonlinear Programming, J. B. Rosen, ed., pp. 447-464. Academic Press, New York.

Barrodale, I., and Roberts, F. D. K. (1973). An improved algorithm for discrete L_1 linear approximation. SIAM J. Numer. Anal. 10: 833-838.

Barrodale, I., and Roberts, F. D. K. (1977). Algorithms for restricted least absolute value estimation. Commun. Stat. Simu. Comput. B6 (4): 353-363.

Bassett, G., Jr., and Koenker, R. (1978). Asymptotic theory of least absolute error regression. J. Am. Stat. Assoc. 73: 618-622.

Bassett, G., Jr., and Koenker, R. (1981). Tests for linear hypotheses and ℓ_1 estimation. Bell Lab. Econ. Discussion Paper 191.

Bloomfield, P., and Steiger, W. (1983). Least Absolute Deviations: Theory, Applications and Algorithms. Birkhäuser, Boston.

Cox, D. R., and Hinkley, D. V. (1974). Theoretical Statistics. Chapman and Hall, London.

Cramer, H. (1946). Mathematical Methods in Statistics. Princeton University Press, Princeton, N. J.

Dielman, T., and Pfaffenberger, R. (1982). LAV (least absolute value) estimation in linear regression: A review. TIMS Stud. Manage. Sci. 19: 31-52.

Ekblom, H., and Henriksson, S. (1969). L_p-criteria for the estimation of location parameters. SIAM J. Appl. Math. 17: 1130-1141.

Forsythe, A. B. (1972). Robust estimation of straight line regression coefficients by minimizing pth power deviations. Technometrics 14: 159-166.

Gentle, J. E. (1977). Least absolute values estimation: An introduction. Commun. Stat. Simul. Comput. B6 (4): 313-328.

Gentle, J. E., Sposito, V. A., and Kennedy, W. J. (1977). On some properties of L_1 estimators. Math. Program. 12: 139-140.

Harter, H. L. (1974-1976). The method of least squares and some alternatives—Parts I-VI. Int. Stat. Rev. 42: 141-174, 159-264, 282 (note added in proof); 43: 1-44, 125-190, 269-278; 44: 111-157.

Harvey, A. C. (1978). On the unbiasedness of robust regression estimators. Commun. Stat. Theor. Meth. A7 (8): 779-783.

Hoaglin, D. C., Mosteller, F., and Tukey, J. W. (1983). Understanding Robust and Exploratory Data Analysis. Wiley, New York.

Hogg, R. V. (1972). More light on the kurtosis and related statistics. J. Am. Stat. Assoc. Theory Methods Sect. 67: 422-424.

Jaeckel, L. A. (1972). Estimating regression coefficients by minimizing the dispersion of the residuals. Ann. Math. Stat. 43: 1449-1458.

Johnson, N. L., and Kotz, S. K. (1970). Continuous Univariate Distributions—I. Houghton Mifflin, Boston.

Josvanger, L. A., and Sposito, V. A. (1983). L_1 norm estimates for the simple regression problem. Commun. Stat. 12: 215-221.

Laplace, P. S. (1793). Sur quelques points du systeme du monde. Mémoires de l'Académie Royale des Sciences de Paris Année 1789, 1-87. Reprinted in Oeuvres Complètes de Laplace, Vol. II, pp. 447-558. Gauthier-Villars, Paris, 1895.

McCormick, G. F., and Sposito, V. A. (1976). Using the L_2 estimator in L_1 estimation. SIAM J. Numer. Anal. 28: 337-343.

Money, A. H., Affleck-Graves, J. F., Hart, H. L., and Barr, G. D. I. (1982). The linear regression model: L_p-norm estimation and the choice of p. Commun. Stat. 11: 89-109.

Nyquist, H. (1980). Recent studies on L_p-norm estimation. Unpublished Ph. D. dissertation, University of Umea, Sweden.

Pfaffenberger, R. C., and Dinkel, J. J. (1978). Absolute deviations curve fitting: An alternative to least squares. Contributions to Survey Sampling and Applied Statistics (Papers in Honor of H. O. Hartley), H. A. David, ed., pp. 279-294. Academic Press, New York.

Rice, J. R., and White, J. S. (1964). Norms for smoothing and estimation. SIAM Rev. 6: 243-256.

Rider, P. R. (1957). The midrange of a sample of an estimator of the population midrange. J. Am. Stat. Assoc. 52: 537-542.

Rosenberg, B., and Carlson, D. (1977). A simple approximation of the sampling distribution of least absolute residuals regression estimates. Commun. Stat. Simul. Comput. B6: 421-428.

Siegel, A. F. (1983). Low median and least absolute residual analysis of two-way tables. J. Am. Stat. Assoc. 78: 371-374.

Sielken, R. L., and Hartley, H. O. (1973). Two linear programming algorithms for unbiased estimation of linear models. J. Am. Stat. Assoc. 68: 639-641.

Sposito, V. A. (1975). Linear and Nonlinear Programming. Iowa State Press, Ames.

Sposito, V. A. (1982). On the unbiasedness of L_p-norm estimators. J. Am. Stat. Assoc. 77: 652-653.

Sposito, V. A., and Tveite, M. D. (1986). On the estimation of the variance of the median used in L_1 linear inference procedures. Commun. Stat. 15 (4): 1367-1375.

Sposito, V. A., Hand, M. L., and McCormick, G. F. (1977). Using an approximate L_1 estimator. Commun. Stat. B6 (3): 263-268.

Sposito, V. A., Hand, M. L., and Skarpness, B. (1983). On the efficiency of using the sample kurtosis in selecting optimal L_p estimators. Commun. Stat. 12: 265-272.

Sposito, V. A., Kennedy, W. J., and Gentle, J. E. (1980). Useful generalized properties of L_1 estimators. Commun. Stat. A9: 1309-1315.

Sposito, V. A., and Smith, W. C. (1976). On a sufficient and a necessary condition for L_1 estimation. Appl. Stat. 25: 154-157.

Sposito, V. A., Smith, W. C., and McCormick, G. F. (1978). Minimizing the sum of absolute deviations. Journal of Applied Statistics and Econometrics, Vandenhoeck and Ruprecht in Gottingen and Zurich Series No. 12, 60 pages.

Tukey, J. W. (1977). Exploratory Data Analyses. Addison-Wesley, Reading, Mass.

Wesolowsky, G. O. (1981). A new descent algorithm for the least value regression problem. Commun. Stat. B10: 479-491.

3

Robust Nonlinear Regression

KENNETH D. LAWRENCE Rutgers University, Piscataway, New Jersey

JEFFREY L. ARTHUR Oregon State University, Corvallis, Oregon

1. INTRODUCTION

The difficulties in fitting a data set to a hypothesized model are well known when the model is linear, especially when there is evidence of one or more outliers in the data. When the model to be fit is nonlinear in the parameters, the problem is further complicated by the numerical difficulties one can encounter in trying to obtain the best estimates for the model's parameters. The purpose of this paper is to present our experiences in dealing with these problems. The paper reports on an experiment involving four small data sets, each of which is to be fit to two nonlinear sigmoidal growth models—namely, a Gompertz model and a logistic model. In addition, the nonlinear models were fit with and without reweighting.

Because nonlinear regression problems are essentially nonlinear programming (NLP) problems, all of the known difficulties in solving NLP problems are present in these situations. One major source of difficulty in any NLP problem is the accuracy of the initial estimates provided to the iterative optimization routine. As a consequence, our experiments also looked at using the L_1 and L_2 norms for providing initial parameter estimates to the nonlinear estimation schemes.

All results were obtained using the SAS statistical software system (SAS Institute, 1985).

2. PRELIMINARIES

2.1 Data Sets

The four data sets used in this experiment have been taken from Ratkowsky (1983) and are presented in Table 1. They deal, respectively, with pasture growth (PASTGR), onion growth (ONION), bean root cells (BEAN), and cucumber cotyledons (CUC). These data have been graphed and are presented in Figures 1 through 4, respectively. From these figures, the sigmoidal nature of these vegetative growth processes is evident.

TABLE 1 Data Sets

Data set 1 (PASTGR)		Data set 2 (ONION)		Data set 3 (CUC)		Data set 4 (BEAN)	
X	Y	X	Y	X	Y	X	Y
9	8.93	1	16.08	0	1.23	.5	1.3
14	10.80	2	33.83	1	1.52	1.5	1.3
21	18.59	3	65.80	2	2.95	2.5	1.9
28	22.33	4	97.20	3	4.34	3.5	3.4
42	39.35	5	191.55	4	5.26	4.5	5.3
57	56.11	6	326.20	5	5.84	5.5	7.1
63	61.73	7	386.87	6	6.21	6.5	10.6
70	64.62	8	520.53	8	6.50	7.5	16.0
79	67.08	9	590.03	10	6.83	8.5	16.4
		10	651.92			9.5	18.3
		11	724.93			10.5	20.9
		12	699.56			11.5	20.5
		13	689.96			12.5	21.3
		14	637.56			13.5	21.2
		15	717.41			14.5	20.9

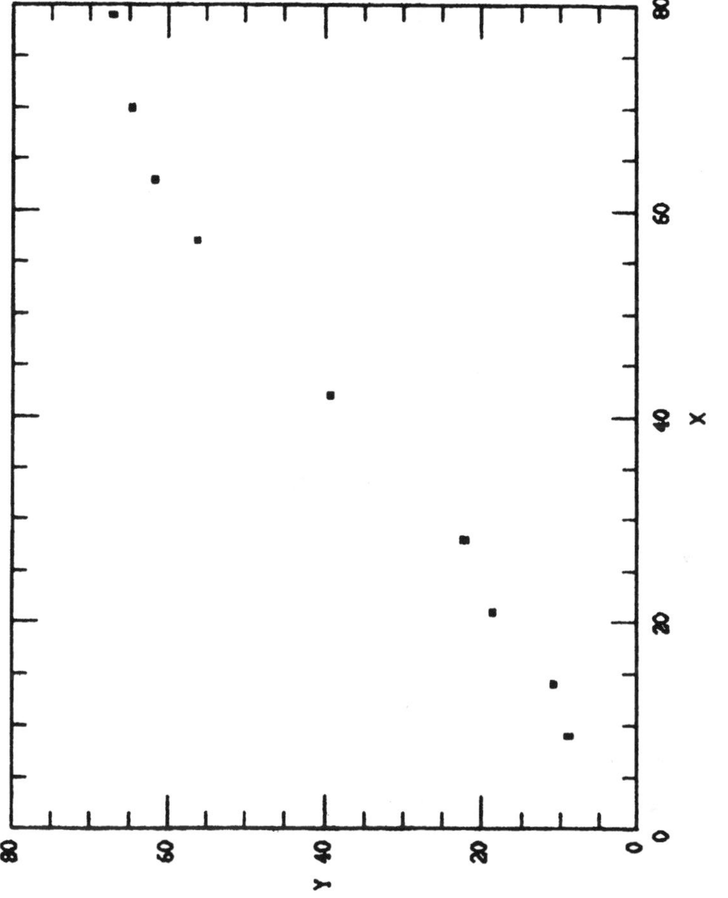

FIGURE 1 The PASTGR data set.

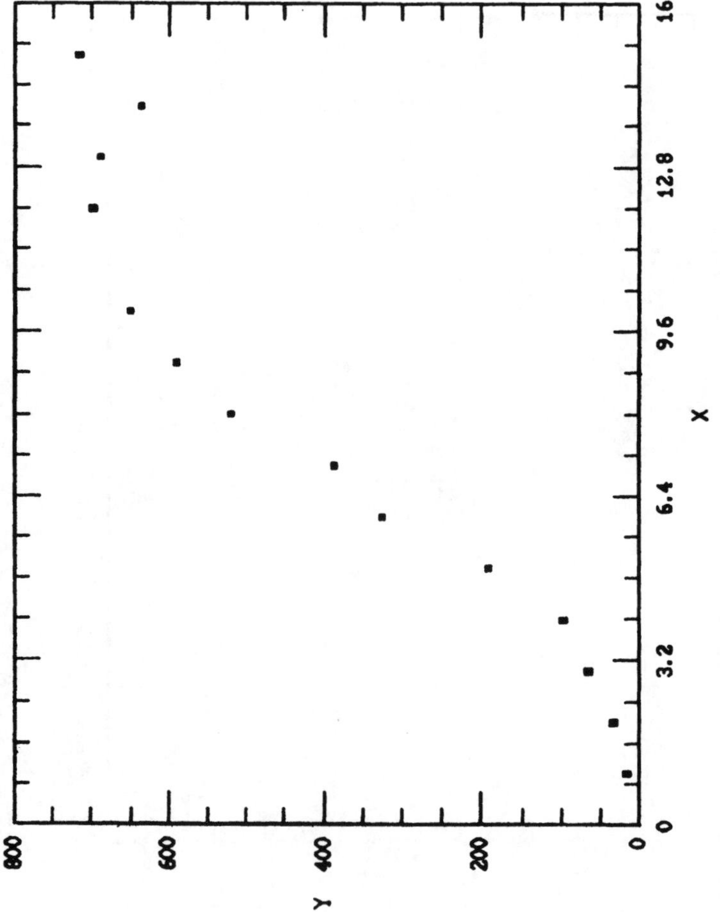

FIGURE 2 The ONION data set.

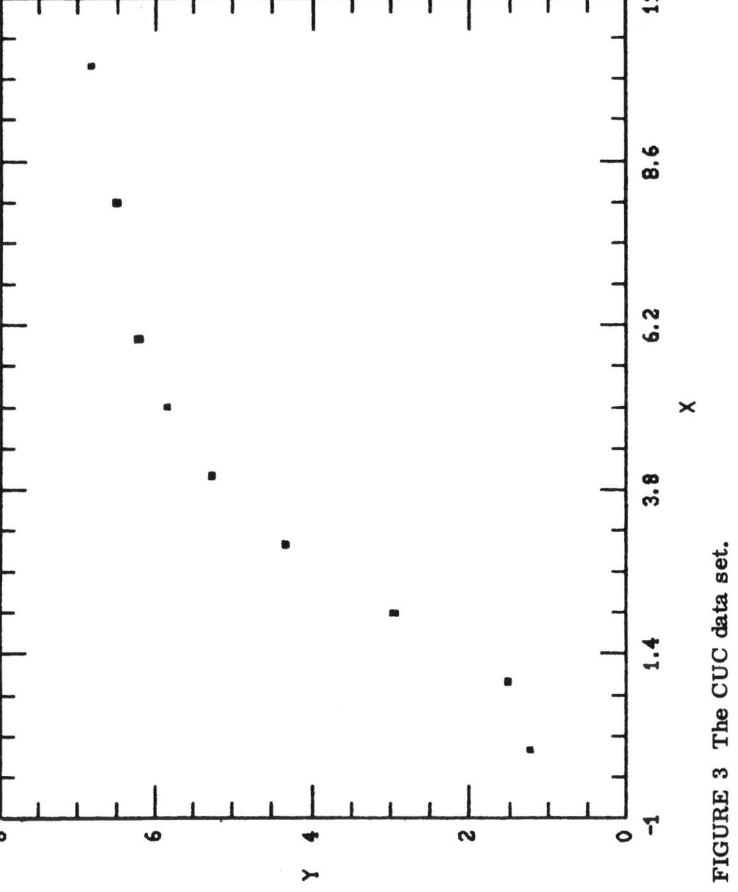

FIGURE 3 The CUC data set.

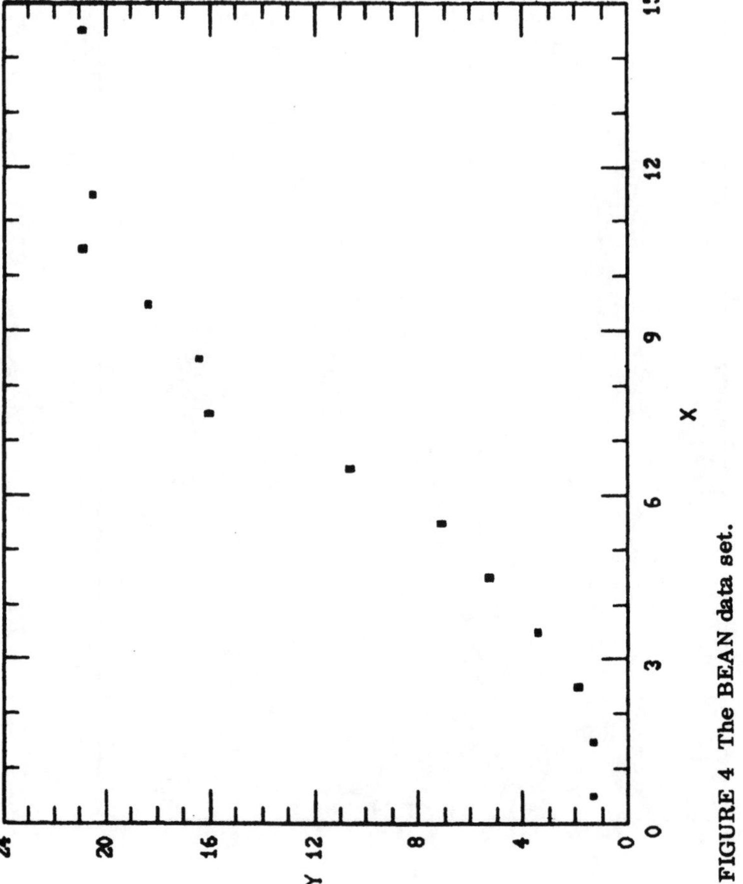

FIGURE 4 The BEAN data set.

2.2 Sigmoidal Growth Models

A number of sigmoidal growth models have been proposed in the literature, many of which have certain underlying theoretical properties. Included are the Gompertz, the logistic, those reported by Richards (1959) and Morgan et al. (1975), and one based on the Weibull distribution. Attention here is restricted to the Gompertz and logistic models.

One form of stating the Gompertz model is provided below:

$$Y = \alpha \exp[-\exp(\beta - \gamma X)] \tag{1}$$

The logistic model is provided in equation (2):

$$Y = \frac{\alpha}{1 + \exp(\beta - \gamma X)} \tag{2}$$

3. INITIAL ESTIMATES

As alluded to earlier, because a nonlinear regression problem is essentially a nonlinear programming problem, the sensitivity of the final solution to the initial estimates must be considered. Previous investigations (see, for example, Arthur 1979) have demonstrated the difficulties encountered due to the use of poor initial parameter estimates.

For the Gompertz model given in equation (1), an initial estimate for α, the asymptote as X approaches ∞, can be found from the graph of the data. For this study, this initial value, denoted α_0, has been selected as the last recorded value of Y, rounded up to the next integer. For the data sets in Table 1, this results in the following values for α_0:

Data set	α_0
PASTGR	68
ONION	718
CUC	7
BEAN	21

Simple algebra on equation (1) results in the following:

$$Z_0 = \log\left[-\log\left(\frac{Y}{\alpha_0}\right)\right] = \beta - \gamma X \tag{3}$$

which is recognized as a simple linear model of X and Z_0. Thus the initial estimates of β and γ, which we shall denote by β_0 and γ_0, respectively, can be obtained via linear regression.

For the logistic model in equation (2), it is again recognized that α represents the asymptotic value of Y as X approaches ∞, so the same values of α_0 were selected as for the Gompertz model above. Manipulation of equation (2) then results in

$$Z_0 = \log\left[\frac{\alpha_0}{Y} - 1\right] = \beta - \gamma X \tag{4}$$

which is again a linear relationship in β and γ. Therefore, initial estimates, β_0 and γ_0, of the parameters β and γ can be obtained once again with simple linear regression.

3.1 The Gompertz Model

Results from applying the least absolute value (LAV) or L_1 norm with the data sets from Table 1 to the linearized Gompertz model in equation (3) are provided in the first column of Table 2. For direct comparisons, the results obtained using the ordinary least squares (OLS) or L_2 norm are provided in the second column of Table 2.

TABLE 2 Initial Estimates for the Gompertz Model

Data set	Parameter	LAV	OLS
PASTGR	β_0	1.55782279	1.72842627
	β_0	0.06180278	0.06718301
ONION	β_0	2.09949533	2.37943152
	β_0	0.40942161	0.46735184
CUC	β_0	0.55325042	0.62319270
	β_0	0.42588499	0.43351717
BEAN	β_0	1.89794752	2.22089156
	β_0	0.40854700	0.51360338

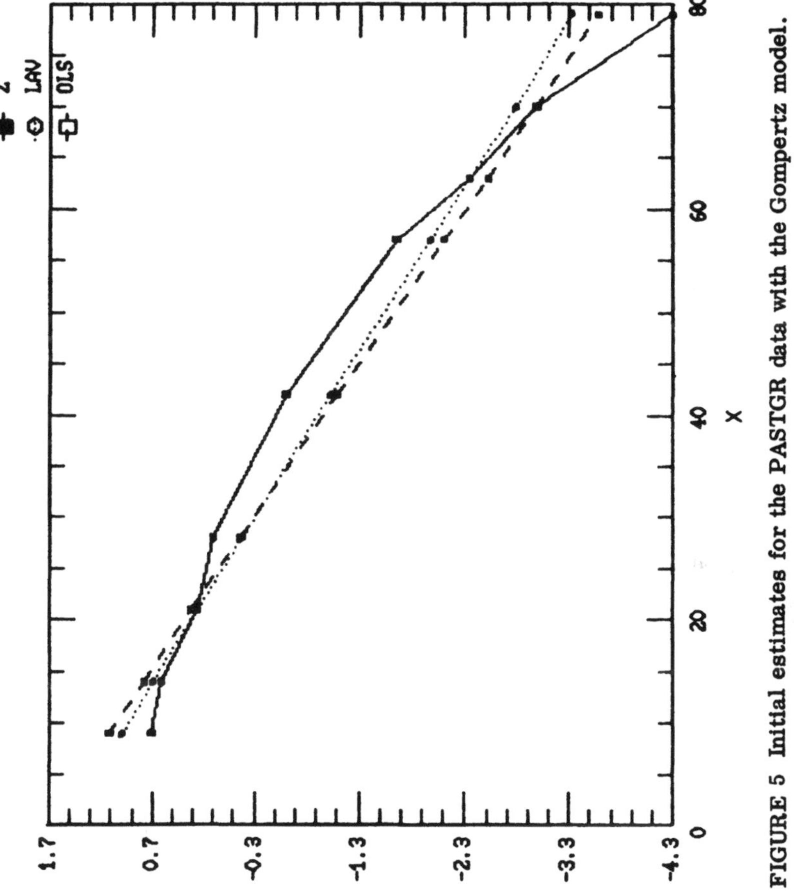

FIGURE 5 Initial estimates for the PASTGR data with the Gompertz model.

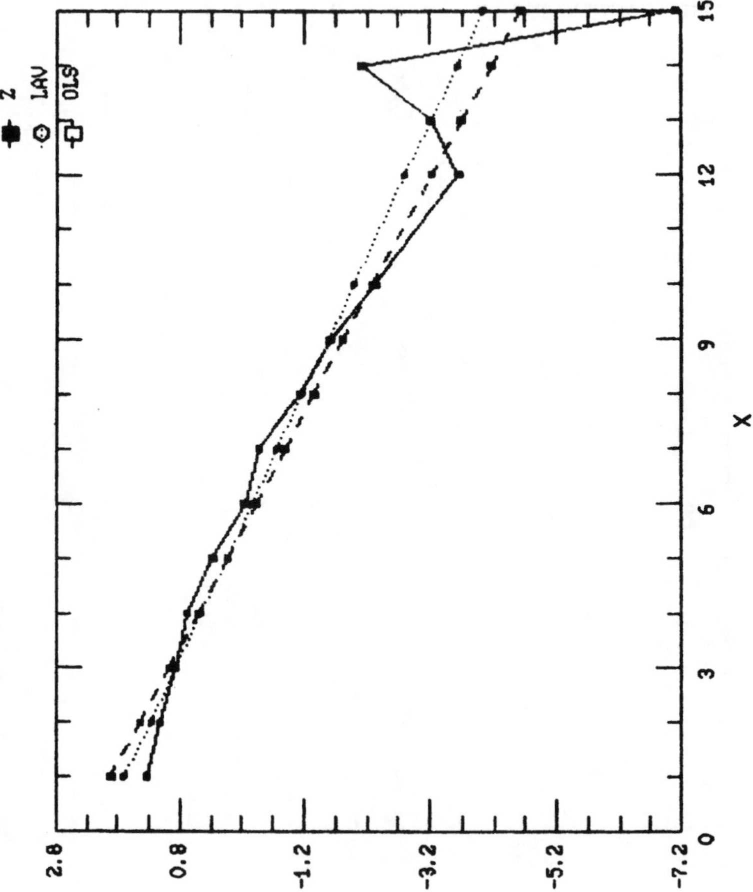

FIGURE 6 Initial estimates for the ONION data with the Gompertz model.

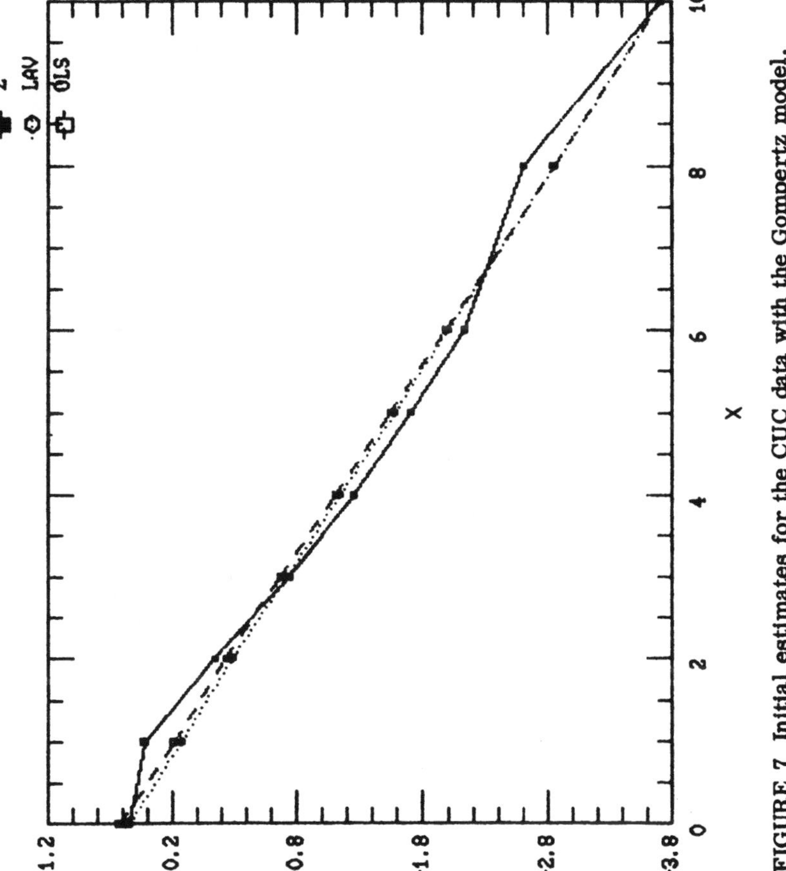

FIGURE 7 Initial estimates for the CUC data with the Gompertz model.

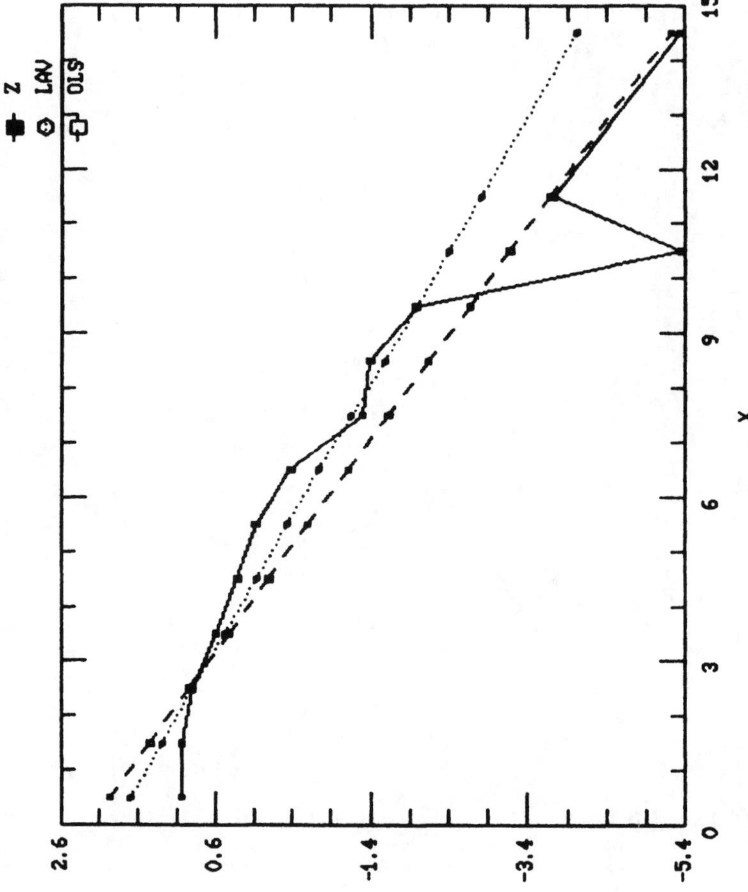

FIGURE 8 Initial estimates for the BEAN data with the Gompertz model.

Graphical results for the four data sets are provided in Figures 5 through 8.

3.2 The Logistic Model

Table 3 contains the results from applying the LAV criteria and the OLS criteria to the linearized logistic model in equation (4) and using the data sets from Table 1. These results are graphed in Figures 9 through 12.

While there are some differences in the estimated values obtained by the LAV and OLS methods, the results produced by the two initial estimation techniques for either nonlinear model are almost exactly the same.

4. NONLINEAR REGRESSION RESULTS

This phase of the experiment was conducted to provide insight into two major questions: (1) Which of the two alternative initial estimation schemes (LAV and OLS) provides better starting points for the fitting of the nonlinear models? (2) What are the effects of allowing reweighting in the fit of the nonlinear models? In particular, the reweighting schemes developed by Hill and Holland (1977), Anscombe (1960), Huber (1981), and Ramsey (1977) were investigated.

TABLE 3 Initial Estimates for the Gompertz Model

Data set	Parameter	LAV	OLS
PASTGR	β_0	2.79672246	2.90269358
	β_0	0.08069391	0.08425785
ONION	β_0	4.30264015	4.18737701
	β_0	0.64789085	0.61883099
CUC	β_0	1.31946002	1.36262834
	β_0	0.50127415	0.52951050
BEAN	β_0	3.71944091	3.78836765
	β_0	0.59296213	0.65743212

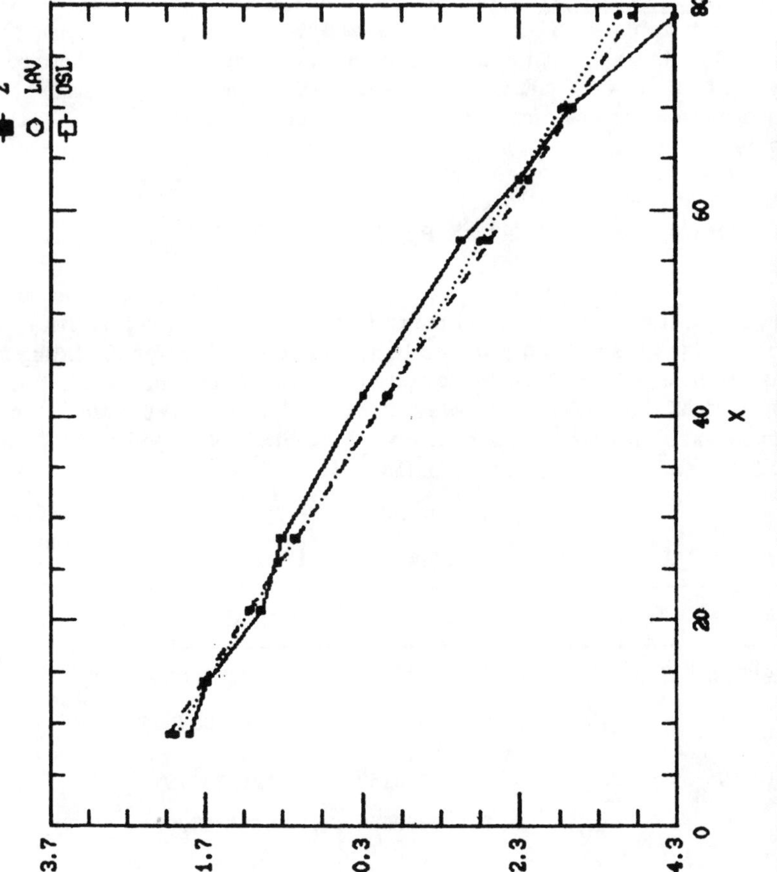

FIGURE 9 Initial estimates for the PASTGR data with the logistic model.

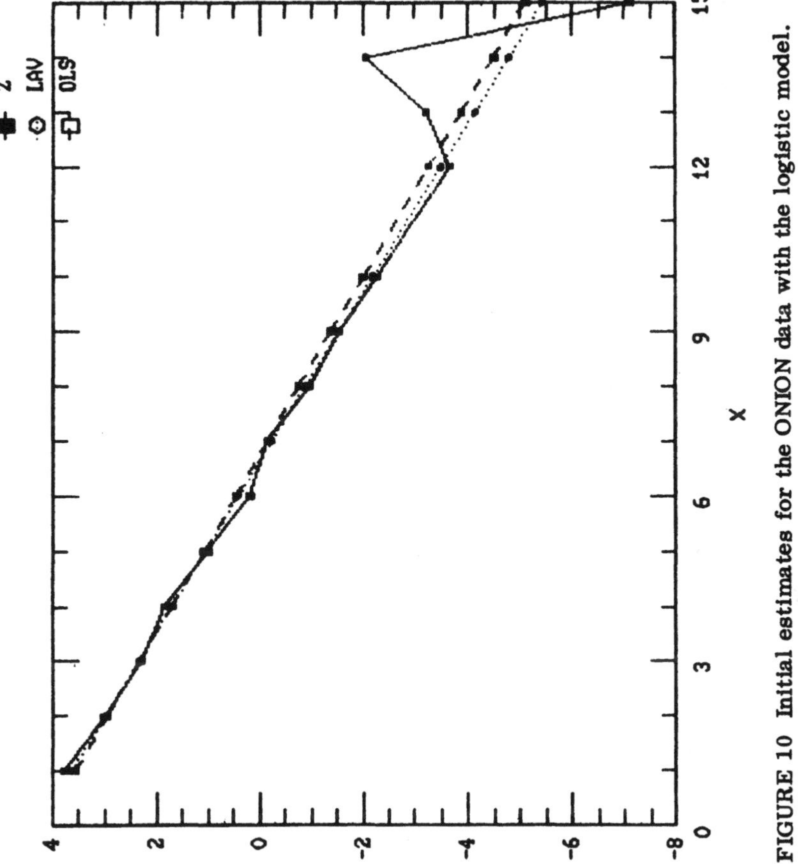

FIGURE 10 Initial estimates for the ONION data with the logistic model.

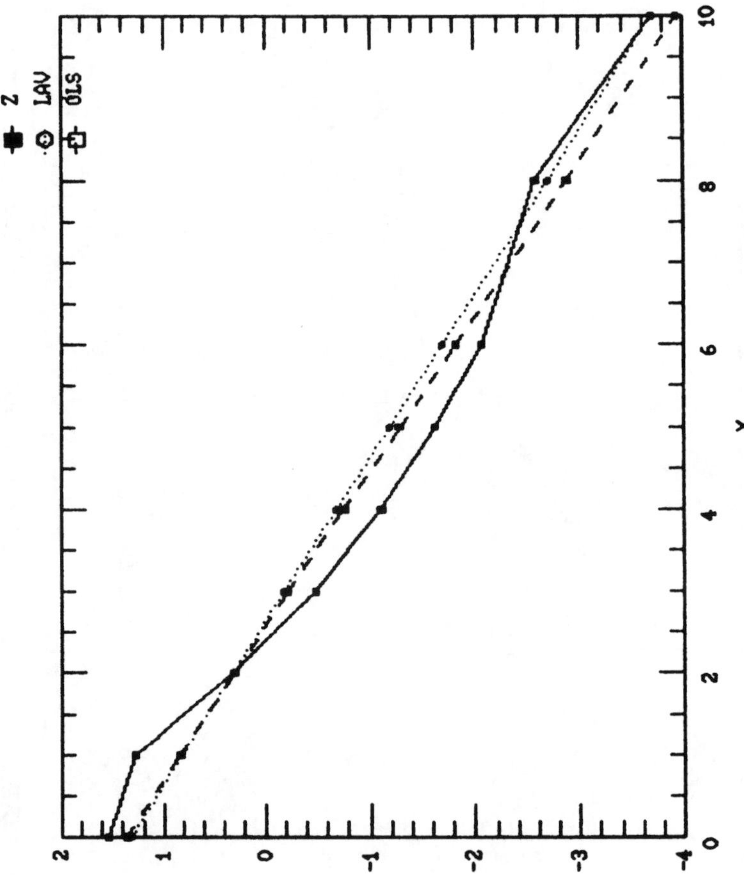

FIGURE 11 Initial estimates for the CUC data with the logistic model.

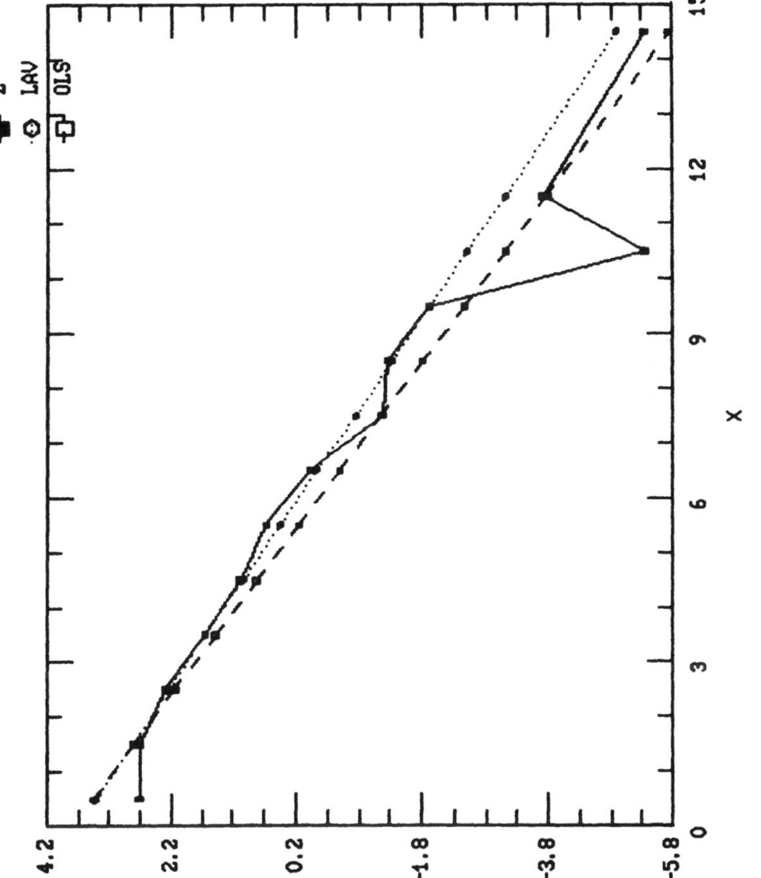

FIGURE 12 Initial estimates for the BEAN data with the logistic model.

The intent in using such weighting functions was to improve the accuracy of the estimation process, as described by Huber (1981) and Montgomery and Peck (1982). Suppose that $\{r_i : i = 1, \ldots, n\}$ represent the residuals obtained by using the nonlinear procedure to fit the model. The Hill-Holland estimator is aone-step iterative reweighting scheme in which a weighted nonlinear least squares regression is performed on the model using weights w_i defined by

$$w_i = \begin{cases} 1 & \text{if } r_i = 0 \\ \dfrac{\sin(r_i/s^*)}{r_i/s^*} & \text{if } 0 < \dfrac{r_i}{s} \leqslant \pi \\ 0 & \text{if } \dfrac{r_i}{s} > \pi \end{cases}$$

Here, s^* is a robust scale estimate obtained as 2.1 times the median of the n - 1 largest $|r_i|$; see Hill and Holland (1977).

The other three reweighting schemes are based on the standardized absolute residuals $e_i = |r_i|/s$, where s is the standard deviation of the residuals $\{r_i\}$. The Anscombe estimator is a one-step iterative reweighting scheme in which a weighted nonlinear least squares regression is performed on the model using weights

$$w_i = \begin{cases} 1 & \text{if } e_i \leqslant k_1 \\ \dfrac{k_1}{e_i} & \text{if } k_1 < e_i \leqslant k_2 \\ 0 & \text{if } e_i > k_2 \end{cases}$$

where k_1 and k_2 are preassigned constants. The Huber estimator uses weights

$$w_i = \begin{cases} 1 & \text{if } e_i \leqslant t \\ \dfrac{t}{e_i} & \text{if } e_i > t \end{cases}$$

for constant t. Finally, the Ramsey estimator uses weights

$$w_i = \exp(-\lambda e_i) \qquad e_i \geqslant 0$$

The preassigned values used for k_1, k_2, t, and λ were chosen as $k_1 = 1.5$, $k_2 = 3.0$, $t = 2.0$, and $\lambda = 0.3$; these values are consistent with those suggested by Montgomery and Peck (1982).

4.1 The Gompertz Model

Results with the Gompertz model are presented in Tables 4 through 7. For each data set and each choice of initial estimates, the tables give the final parameter estimates and the residual sum of squares (RSSE). Several observations can be made from an informal perusal of these results. First, there appears to be almost no difference in the results obtained by using the L_2 initial estimates when compared with the L_1 initial estimates. Second, the unweighted nonlinear regression scheme performs admirably, obtaining the best estimates (in terms of the RSSE) for both the PASTGR and CUC data sets; in fact, the results were identical to those found by the reweighting schemes of Huber and Anscombe. Third, the reweighting method developed by Huber performed best on two data sets (ONION and CUC). Finally, the method of Ramsey was consistently outperformed by at least one of the other procedures.

4.2 The Logistic Model

Analogous results for the logistic model are presented in Tables 8 through 11. For the most part, the conclusions reached with the Gompertz model are applicable with regard to these results. In particular, the choice of initial estimation schemes does not seem to make a difference in the final estimates obtained. Again, the unweighted nonlinear regression procedure performed well, actually obtaining the best estimates (again in terms of the RSSE) with all four data sets. The reweighting methods proposed by Anscombe and Huber also performed well, obtaining minimal RSSE in three out of the four cases. Finally, the procedure of Ramsey was again outperformed (and in an even more significant way) by the other procedures.

TABLE 4 Results with the Gompertz Model and the PASTGR Data Set

Regression scheme		LAV initial estimates	OLS initial estimates
Unweighted	$\hat{\alpha}$	82.8324	82.8318
	$\hat{\beta}$	0.0371	0.0371
	$\hat{\gamma}$	1.2237	1.2237
	RSSE	21.7940	21.7940
Hill-Holland	$\hat{\alpha}$	77.5071	77.5083
	$\hat{\beta}$	0.0384	0.0384
	$\hat{\gamma}$	1.2402	1.2402
	RSSE	47.4542	47.4473
Anscombe	$\hat{\alpha}$	82.8324	82.8324
	$\hat{\beta}$	0.0371	0.0371
	$\hat{\gamma}$	1.2237	1.2237
	RSSE	21.7940	21.7940
Huber	α	82.8324	82.8318
	$\hat{\beta}$	0.0371	0.0371
	$\hat{\gamma}$	1.2237	1.2237
	RSSE	21.7940	21.7940
Ramsey	$\hat{\alpha}$	70.8000	70.7553
	$\hat{\beta}$	0.0390	0.0383
	$\hat{\gamma}$	1.2386	1.2388
	RSSE	42.8174	42.8270

TABLE 5 Results with the Gompertz Model and the ONION Data Set

Regression scheme		LAV initial estimates	OLS initial estimates
Unweighted	$\hat{\alpha}$	723.101	723.118
	$\hat{\beta}$	2.500	2.500
	$\hat{\gamma}$	0.450	0.450
	RSSE	13606.1	13606.1
Hill-Holland	$\hat{\alpha}$	721.405	721.350
	$\hat{\beta}$	2.260	2.261
	$\hat{\gamma}$	0.416	0.416
	RSSE	14368.2	14366.4
Anscombe	$\hat{\alpha}$	729.057	729.148
	$\hat{\beta}$	2.147	2.146
	$\hat{\gamma}$	0.391	0.391
	RSSE	15210.0	15218.4
Huber	$\hat{\alpha}$	717.097	716.756
	$\hat{\beta}$	2.504	2.510
	$\hat{\gamma}$	0.454	0.455
	RSSE	11943.1	13713.8
Ramsey	$\hat{\alpha}$	700.194	700.225
	$\hat{\beta}$	1.979	1.979
	$\hat{\gamma}$	0.390	0.390
	RSSE	26301.2	26291.9

TABLE 6 Results with the Gompertz Model and the CUC Data Set

Regression scheme		LAV initial estimates	OLS initial estimates
Unweighted	$\hat{\alpha}$	6.9241	6.9250
	$\hat{\beta}$	0.7685	0.7680
	$\hat{\gamma}$	0.4936	0.4933
	RSSE	0.3717	0.3717
Hill-Holland	$\hat{\alpha}$	6.8320	6.8329
	$\hat{\beta}$	0.8863	0.8853
		0.5447	0.5442
	RSSE	0.3921	0.3917
Anscombe	$\hat{\alpha}$	6.9249	6.9249
	$\hat{\beta}$	0.7681	0.7681
	$\hat{\gamma}$	0.4933	0.4933
	RSSE	0.3717	0.3717
Huber	$\hat{\alpha}$	6.9249	6.9249
	$\hat{\beta}$	0.7680	0.7681
	$\hat{\gamma}$	0.4933	0.4933
	RSSE	0.3717	0.3717
Ramsey	$\hat{\alpha}$	6.6087	6.6123
	$\hat{\beta}$	0.8592	0.8583
	$\hat{\gamma}$	0.5567	0.5558
	RSSE	0.5274	0.5237

TABLE 7 Results with the Gompertz Model and the BEAN Data Set

Regression scheme		LAV initial estimates	OLS initial estimates
Unweighted	$\hat{\alpha}$	22.5074	22.5048
	$\hat{\beta}$	2.1058	2.1077
	$\hat{\gamma}$	0.3880	0.3883
	RSSE	12.5905	12.5095
Hill-Holland	$\hat{\alpha}$	21.9742	21.9770
	$\hat{\beta}$	2.1188	2.1188
	$\hat{\gamma}$	0.3990	0.3989
	RSSE	8.9730	8.6138
Anscombe	$\hat{\alpha}$	22.5067	22.5065
	$\hat{\beta}$	2.1063	2.1065
	$\hat{\gamma}$	0.3881	0.3881
	RSSE	12.5905	12.5905
Huber	$\hat{\alpha}$	22.5067	22.5065
	$\hat{\beta}$	2.1063	2.1065
	$\hat{\gamma}$	0.3881	0.3881
	RSSE	12.5905	12.5905
Ramsey	$\hat{\alpha}$	20.8282	20.8323
	$\hat{\beta}$	2.0843	2.0845
	$\hat{\gamma}$	0.4193	0.4192
	RSSE	14.0021	13.8639

TABLE 8 Results with the Logistic Model and the PASTGR Data Set

Regression scheme		LAV initial estimates	OLS initial estimates
Unweighted	$\hat{\alpha}$	72.4622	72.4615
	$\hat{\beta}$	2.6180	2.6181
	$\hat{\gamma}$	0.0674	0.0674
	RSSE	8.0565	8.0565
Hill-Holland	$\hat{\alpha}$	70.3465	70.3476
	$\hat{\beta}$	2.6195	2.6195
	$\hat{\gamma}$	0.0702	0.0702
	RSSE	10.5832	10.5803
Anscombe	$\hat{\alpha}$	72.4622	72.4622
	$\hat{\beta}$	2.6181	2.6180
	$\hat{\gamma}$	0.0674	0.0674
	RSSE	8.0565	8.0565
Huber	$\hat{\alpha}$	72.4622	72.4622
	$\hat{\beta}$	2.6181	2.6180
	$\hat{\gamma}$	0.0674	0.0674
	RSSE	8.0565	8.0565
Ramsey	$\hat{\alpha}$	66.5868	66.5863
	$\hat{\beta}$	2.6005	2.6005
	$\hat{\gamma}$	0.0744	0.0744
	RSSE	36.4268	36.4302

TABLE 9 Results with the Logistic Model and the ONION Data Set

Regression scheme		LAV initial estimates	OLS initial estimates
Unweighted	$\hat{\alpha}$	702.844	702.864
	$\hat{\beta}$	4.444	4.443
	$\hat{\gamma}$	0.688	0.689
	RSSE	8929.9	8929.9
Hill-Holland	$\hat{\alpha}$	688.258	688.260
	$\hat{\beta}$	3.522	3.522
	$\hat{\gamma}$	0.579	0.579
	RSSE	20200.8	20014.6
Anscombe	$\hat{\alpha}$	718.551	718.471
	$\hat{\beta}$	4.099	4.103
	$\hat{\gamma}$	0.626	0.626
	RSSE	10246.8	10224.1
Huber	$\hat{\alpha}$	691.783	692.182
	$\hat{\beta}$	4.001	3.995
	$\hat{\gamma}$	0.627	0.626
	RSSE	12057.5	12072.4
Ramsey	$\hat{\alpha}$	624.420	624.165
	$\hat{\beta}$	3.790	3.791
	$\hat{\gamma}$	0.663	0.664
	RSSE	48429.1	48619.3

TABLE 10 Results with the Logistics Model and the CUC Data Set

Regression scheme		LAV initial estimates	OLS initial estimates
Unweighted	$\hat{\alpha}$	6.6868	6.6867
	$\hat{\beta}$	1.7451	1.7451
	$\hat{\gamma}$	0.7547	0.7547
	RSSE	0.2119	0.2119
Hill-Holland	$\hat{\alpha}$	6.4733	6.4733
	$\hat{\beta}$	1.8819	1.8819
	$\hat{\gamma}$	0.8500	0.8500
	RSSE	0.3077	0.3077
Anscombe	$\hat{\alpha}$	6.6868	6.6867
	$\hat{\beta}$	1.7452	1.7451
	$\hat{\gamma}$	0.7547	0.7547
	RSSE	0.2119	0.2119
Huber	$\hat{\alpha}$	6.6868	6.6867
	$\hat{\beta}$	1.7452	1.7451
	$\hat{\gamma}$	0.7547	0.7547
	RSSE	0.2119	0.2119
Ramsey	$\hat{\alpha}$	5.8649	5.8648
	$\hat{\beta}$	1.9717	1.9718
	$\hat{\gamma}$	1.0038	1.0039
	RSSE	1.8722	1.8723

TABLE 11 Results with the Logistic Model and the BEAN Data Set

Regression scheme		LAV initial estimates	OLS initial estimates
Unweighted	$\hat{\alpha}$	21.5089	21.5093
	$\hat{\beta}$	3.9574	3.9572
	$\hat{\gamma}$	0.6222	0.6221
	RSSE	6.2102	6.2102
Hill-Holland	$\hat{\alpha}$	21.3557	21.3555
	$\hat{\beta}$	3.4273	3.4274
	$\hat{\gamma}$	0.5498	0.5498
	RSSE	8.5167	8.5154
Anscombe	$\hat{\alpha}$	21.5090	21.5088
	$\hat{\beta}$	3.9573	3.9573
	$\hat{\gamma}$	0.6221	0.6222
	RSSE	6.2102	6.2102
Huber	$\hat{\alpha}$	21.5090	21.5088
	$\hat{\beta}$	3.9573	3.9573
	$\hat{\gamma}$	0.6221	0.6222
	RSSE	6.2102	6.2102
Ramsey	$\hat{\alpha}$	20.0895	20.0889
	$\hat{\beta}$	3.4444	3.4445
	$\hat{\gamma}$	0.5922	0.5922
	RSSE	19.7347	19.7393

5. CONCLUSIONS

Results of a computational experiment with various nonlinear regression procedures have been described. In particular, the experiment has used two different sigmoidal growth models, four different data sets, two initial estimation schemes, and five nonlinear estimation procedures (four of which involve reweighting). The results show, among other things, that the choice of schemes for obtaining initial estimates had no influence on the final estimates and that the performance of the nonlinear regression procedure with no reweighting was remarkably close to that of the more complicated procedures that reweight.

Future experiments would be useful in examining the effect of one or more serious outliers on the performance of these various reweighting schemes.

REFERENCES

Anscombe, F. J. (1960). Rejection of outliers. Technometrics 2: 123-147.

Arthur, J. L. (1979). A comparison of optimization methods on a nonlinear regression problem. Tech. Rep. No. 70, Dept. of Statistics, Oregon State University, Corvallis.

Hill, R. W., and Holland, T. W. (1977). Two robust alternatives to least squares regression. J. Am. Stat. Assoc. 72: 828-833.

Huber, P. J. (1981). Robust Statistics. Wiley, New York.

Montgomery, D. C., and Peck, E. A. (1982). Introduction to Linear Regression Analysis. Wiley, New York.

Morgan, P. H., Mercer, L. P., and Flodin, N. W. (1975). General model for nutritional responses of higher organisms. Proc. Natl. Acad. Sci. U.S.A. 72: 4327-4331.

Ramsey, J. O. (1977). A comparative study of several robust estimates of slope, intercept and scale in linear regression. J. Am. Stat. Assoc. 72: 608-615.

Ratkowsky, D. A. (1983). Nonlinear Regression Modeling: A Unified Practical Approach. Marcel Dekker, New York.

Richards, F. J. (1959). A flexible growth function for empirical use. J. Exp. Biol. 10: 290-300.

SAS Institute, Inc. (1985). SAS User's Guide: Statistics. Cary, N.C.

Part II

Robust Regression Methods

4

An Algorithm to Assist in the Identification of Multiple Multivariate Outliers When Using a Least Absolute Value Criterion

RONALD D. ARMSTRONG Rutgers University, Newark, New Jersey

PHILIP O. BECK American Airlines, Dallas, Texas

1. INTRODUCTION

Least absolute value (LAV), or L_1 norm, estimates have long been recognized as an acceptable alternative to least square estimates. Fourier appears to have been the first to consider the problem of obtaining such estimates (Charnes et al., 1955), but until the comparatively recent development of the electronic digital computer and linear programming the labor involved in LAV estimation made the technique unpopular. Also, when the assumption of normality of the random variable is made, strong theoretical justification can be stated for least square estimates (Koenker and Bassett, 1978). The importance of LAV estimates in multiple linear regression analysis is becoming widely accepted in the statistical community. The two major factors contributing to the popularity of LAV regression are the sensitivity of least squares to outliers and the inappropriateness of a Gaussian error assumption. LAV is particularly well suited to problems where the error distribution is fat-tailed (e.g., Laplace or Cauchy distributed). Surveys of LAV algorithmic procedures, statistical properties, and empirical tests may be found in Armstrong and Kung (1982a), Dielman and Pfaffenberger (1982), Gentle (1977), Gentle et al. (1977), Kennedy and Gentle (1980), and Koenker and Bassett (1978).

The classical linear regression model can be stated as follows:

$$Y = \sum_{j=1}^{m} X_j \beta_j + \epsilon$$

where Y is the dependent random variable; X_j, j = 1, 2, ..., m, are the independent random variables; β_j, j = 1, 2, ..., m, are the parameters of the model; and ϵ is an error term. A problem arises when n sets of values are observed for ($Y, X_1, X_2, \ldots, X_m$) and β must be estimated. The classical approach is to choose β in order to minimize the sum of the error terms in a specified sense. The criterion considered here is LAV and the problem becomes

$$\text{minimize} \sum_{i \in N} \left| y_i - \sum_{j=1}^{m} x_{ij} \beta_j \right| \tag{1}$$

where y_i and X_{ij} are the given observations and $N = \{1, 2, \ldots, n\}$, n being the number of observations.

Barnett and Lewis (1980) discussed the importance of analyzing and detecting outliers in statistical data. Outliers may arise because of the inherent variability in which observations vary over the population, or errors in measuring members of a population under study, or errors in the collection of data.

Mickey (1974) and Mickey et al. (1967) consider identifying outliers in least squares regression. Their approach pinpoints an observation as an outlier if the deletion of that observation results in a large reduction in the sum of squares of residuals. This approach involves the fitting of n least squares regressions, each computed with one of the n observations deleted from the data set.

Gentleman and Wilk (1975) propose methods for identifying the subset of k most likely outliers in least squares regression. They use analysis of variance methods to identify the subset of k most likely outliers by computing a sum of squares associated with the omitted k data subset.

This chapter proposes a method for determining multiple outliers in least absolute value regression. This method is similar to the procedure described by Gentleman and Wilk (1975) in that subsets of outliers are determined.

The method identifies k potential outliers for k = 1, 2, ..., K by considering the simultaneous deletion of k observations from the set of n observations and performing a regression analysis on each of the reduced samples. So a total of $\Sigma_{k=1}^{K} \binom{n}{k}$ least absolute value regression problems is considered. The LAV version of the problem can be stated formally as follows. Determine the values of I and β which

$$\text{minimize} \sum_{i \epsilon I} \left| y_i - \sum_{j=1}^{m} x_{ij}\beta_j \right| \tag{2}$$

$N \supset I$ and the cardinality of I equals n, n - 1, ..., n - k. In other words, for k = 0, 1, ..., K determine the k observations whose removal from the problem will reduce the sum of absolute deviations the most. A total of K + 1 different β vectors and index sets I will be obtained.

2. ALGORITHMIC FRAMEWORK

The algorithm used to solve the best subset problem is an implicit enumeration procedure that solves a problem of the form given by (2) at each stage. Rewriting (2) in a linear programming equivalent form yields

$$\text{minimize } z = \sum_{i \epsilon I} (d_i^+ + d_i^-) \tag{3}$$

subject to

$$\sum_{j=1}^{m} x_{ij}\beta_j + d_i^+ - d_i^- = y_i, \quad d_i^+ \geqslant 0,\ d_i^- \geqslant 0,\ i \,\epsilon\, I$$

where d_i^+ and d_i^- are positive and negative deviations of the ith observation.

The linear programming dual of (3) is the following.

$$\text{maximize } w = \sum_{i \epsilon I} y_i \pi_i \tag{4}$$

subject to

$$\sum_{i \epsilon I} x_{ij}\pi_i = 0, \quad j = 1, 2, \ldots, m, \ -1 \leqslant \pi_i \leqslant +1,\ i \,\epsilon\, I$$

It is easily shown that (3) and (4) will always have optimal objective values that are equal. Also, the simplex algorithm for linear programming problems will readily provide the optimal π values for (4) once (3) is solved and, similarly, the optimal β values are available once (4) is solved. Thus, computational considerations alone should determine whether (3) or (4) should be solved with the simplex algorithm.

Special-purpose simplex algorithms have been developed for both (3) and (4). A simplex algorithm that maintains a feasible solution to (3) at each iteration will be termed a dual algorithm. The study in Armstrong and Kung (1982a) indicates that a variant of the Barrodale and Roberts (1973) procedure which is a primal algorithm is more efficient than a specialized dual algorithm. This result is, however, based on solving a single LAV problem with no previous information. The algorithm to be proposed for identifying outlying observations repeatedly considers LAV problems where previous information can be used to restart the procedure to solve an individual LAV problem. Results for determining the best subset of parameters in LAV (see Armstrong and Kung, 1982b) showed that the dual algorithm was more efficient.

An algorithm to assist in identifying outliers will now be stated which is independent of the method used to solve (2).

2.1 Variable Definitions

$X_i = (x_{i1}, x_{i2}, \ldots, x_{im})$: The ith row of the observation matrix X.
q: The number of observations in the current subproblem.
SAD_i: The current best objective with i observations.
$STAT_i = 0$: ith observation is in the model but has not been forced in.
$STAT_i = 1$: ith observation is forced in the model.
$STAT_i = -1$: ith observation is forced out of the model.
ℓ: The current level in the solution tree. The initial problem is at level zero and a node is one level deeper in the tree than the immediate predecessor.
PAR_ℓ: The observation restricted at level ℓ of the predecessor path. If PAR_ℓ is negative the observation is forced out of model and if PAR_ℓ is positive the observation is forced in the model.

One trivial modification on the algorithm is to force an X_r to be included in every regression. This can be accomplished by setting $STAT_r$ equal to 1 rather than 0 at STEP 1.

2.2 Steps of the Algorithm

STEP 1. Set $q = n$; $SAD_1 = \infty$, $i = n - k, \ldots, n$, $\ell = 0$, $I = \{1, 2, \ldots, n\}$, and $STAT_i = 0$, $i = 1, 2, \ldots, n$.

STEP 2. Solve (2) to obtain an optimal solution $(\bar{z}, \bar{\beta})$. If $\bar{z} > SAD_q$, then go to STEP 4; otherwise, go to STEP 3.

STEP 3. A better solution has been found for a subset with q observations included in the model. Set $SAD_q = \bar{z}$ and save $\bar{\beta}$.

STEP 4. If $q \leftarrow n - K$ then go to STEP 6; otherwise, set $q \leftarrow q - 1$ and $\ell \leftarrow \ell + 1$.

STEP 5. Choose an observation X_u with $STAT_u = 0$ to force out of the model. (See Section 4 for a discussion of branching procedures.) Form a new subproblem with $\pi_u = 0$, $STAT_u = 1$, and $PAR_\ell = -u$, and remove u from I. Go to STEP 2.

STEP 6. If $PAR_\ell > 0$ then go to STEP 7; otherwise, set $PAR_\ell \leftarrow -PAR_\ell$, $i = PAR_\ell$, $STAT_i = +1$, and $q \leftarrow q + 1$. Go to STEP 4.

STEP 7. Set $i = PAR_\ell$, $STAT_i = 0$, and $\ell \leftarrow \ell - 1$. If $\ell > 0$ then go to STEP 6; otherwise, terminate the enumeration process.

The optimal solution of (2) is not used when $\bar{z} > SAD_q$; thus, it is not necessary to solve (2) if it can be ascertained that $\bar{z} > SAD_q$ by another test.

An additional test is easily implemented when (4) is solved using the primal simplex algorithm. The algorithm will maintain a feasible solution to (4) and the objective value (w) will be monotonically nondecreasing from iteration to iteration. Therefore the solution procedure can be terminated whenever the following holds:

$$w > SAD_q \tag{5}$$

The simplex algorithm applied to (3) will maintain a feasible solution to (3) and the objective will be monotonically nonincreasing from iteration to iteration. Thus, the optimal solution to (3), $\bar{z}$, must generally be obtained before it can be determined whether or not this subset yields a better regression than any previously considered model with exactly q observations. As noted by Armstrong and Kung (1982b), the early termination test allowed by solving (4) directly leads to significant computational savings in an analogous best parameter subset problem. The branch-and-bound procedure of this chapter will test both the primal and dual algorithms to solve the LAV problems.

3. ADVANCED STARTS AND BOUNDING RULES

The previously described branch-and-bound algorithm uses a last-in-first-out (LIFO) branching rule. Viewing the algorithm in a tree format, every node corresponds to a linear programming problem. A complete tree with n = 5 and K = 2 is given in Figure 1. Two problems are formed by considering the "current" problem, removing an abservation from the model on one branch, and forcing the same observation to be included in the model on the other branch. Once a condition is specified, it must be satisfied in all descendants of the node.

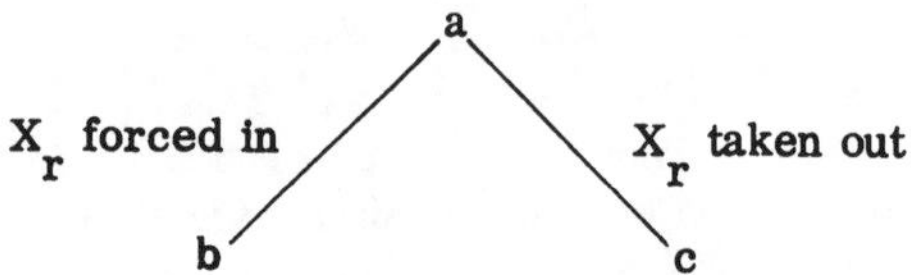

The branch where X_r is forced in the model gives rise to the same linear programming problem in the immediate predecessor node. Thus, the problem at node b of the diagram need not be solved. The problem of concern arises when some X_r is removed from the model. Let (4) represent the problem at node a. Setting $\pi_r = 0$ in (3) is the same as removing the rth observation from (4). The problem at node c written by modifying the problem at node a can be stated in the dual form as follows.

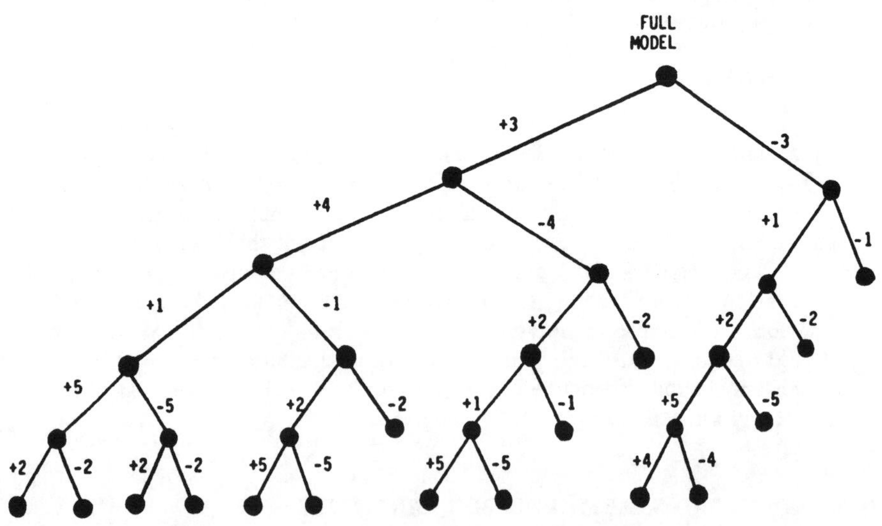

FIGURE 1 A complete solution tree is depicted when n = 2 and K = 2. A -i indicates the removal of X_i and +i indicates the forced inclusion of X_i

$$\text{maximize } w = \sum_{\substack{i \varepsilon I \\ j \varepsilon \{I-r\}}}^{n} y_i \pi_i = 0 \qquad j = 1, 2, \ldots, m \tag{6}$$

$$-1 \leqslant \pi_i \leqslant +1 \qquad i \,\varepsilon \,\{i - r\}$$

$$\pi_i = 0 \qquad \text{otherwise}$$

It will be assumed that the matrix (x_{ij}) has rank m. This assumption is made to facilitate the following presentation, and rank deficiencies are easily handled with the specialized linear programming algorightms. From this assumption it follows that the primal and dual algorithms utilize a working basis of size m × m. At any stage of either the primal or dual algorithm the following definitions apply.

B: An m × m basis matrix for the linear programming problem at node a. This basis is formed from the coefficient matrix (x_{ij}).
B^{-1}: The inverse of B.
b_{ij}: The (i, j) entry in B^{-1}.
$\overline{x}_{ik}$: $\Sigma_{j=1}^{m} b_{ij}x_{jk}$.
IB = $\{k(1), k(2), \ldots, k(m)\}$: An ordered index set defining the π variables in the basis.
π: The current values of the π vector. Note that π_i = +1 or -1 for i ε IB. $\pi_i = 0$ for i ε IB implies that X_i has been removed from the model.
$(\pi_{k(1)}, \overline{\pi}_{k(2)}, \ldots, \overline{\pi}_{k(m)})$: The vector of basic $\overline{\pi}$ variables. $\overline{\pi}_{k(i)} = -\Sigma_{j=1}^{m} b_{ij}(\Sigma_{i \varepsilon IB} x_{ij}\overline{\pi}_i)$.

The process of considering problem (6) will now be discussed from both the primal and dual viewpoints.

3.1 Dual Algorithm

The removal of an observation from the model is equivalent to removing a column from the dual linear programming problem. The additional restriction $\pi_r = 0$ is added and the dual solution of the immediate predecessor becomes infeasible unless π_r happens to be basic with a value of zero. Two major cases arise.

Case 1. The restricted variable π_r is nonbasic in the final solution to the immediate predecessor.

The value of the π vector is updated as follows.

$$\bar{\pi}_{k(i)} \leftarrow \bar{\pi}_{k(i)} + \bar{x}_{ir}\bar{\pi}_r \qquad i = 1, 2, \ldots, m$$

$$\bar{\pi}_r = 0$$

All other π_j remain unchanged.

If $-1 \leqslant \bar{\pi}_{k(i)} \leqslant +1$, $i = 1, 2, \ldots, m$, the updated solution is feasible. Furthermore, if the optimal solution to the immediate predecessor was obtained, then the updated solution is optimal for this subproblem. If the last solution to the immediate predecessor was not optimal but the updated dual solution is feasible, then the algorithm can proceed directly to phase 2 of the dual method; otherwise, a phase 1 routine is employed to obtain dual feasibility.

Case 2. The restricted variable π_r is basic in the final solution to the immediate predecessor.

The variable π_r must be removed from the basis and forced to zero. This can be accomplished by replacing π_r in the basis with a nonbasic variable π_s such that

$$\text{sign}(\bar{\pi}_r) = \text{sign}(-\bar{\pi}_s\bar{x}_{ps})$$

where $k(p) = r$.

After π_s enters the basis and all appropriate updates have taken place, the algorithm checks for feasibility of the basic variables. A phase 2 routine is entered immediately if dual feasibility is obtained and a phase 1 routine is entered otherwise.

A major advantage of the dual algorithm is that, once phase 2 is entered, the objective value is monotonically increasing. This means that the solution of the current linear program can be terminated prematurely whenever $\bar{w} > \bar{Z}$.

3.2 Primal Algorithm

The removal of an observation from the model is equivalent to removing a constraint from the primal linear programming problem. Thus, primal feasibility is maintained but the optimality conditions (dual feasibility) may be lost. The same two major cases that arose with the dual algorithm are also found here.

Case 1. The restricted variable π_r is nonbasic in the final solution to the immediate predecessor.

The value of the π vector can be updated as described in Case 1 of the dual algorithm and the primal algorithm can be entered immediately after the update of all appropriate indicators.

Case 2. The restricted variable is basic in the final solution to the immediate predecessor.

The variable π_r is removed from the basis and forced to zero. The variable to enter the basis must be chosen to maintain primal feasibility. Thus, unlike the dual algorithm approach, a minimum ratio procedure is employed. After π_r has been replaced in the basis and the appropriate updates have taken place, the primal algorithm for LAV regression problems can begin.

4. COMPUTATION RESULTS

Primal and dual implementations of the algorithm for identifying outlying observations have been coded in standard FORTRAN. Regression problems from the literature and randomly generated problems were solved. To facilitate the analysis of dimension modification and to permit a tabular presentation, only results with randomly generated problems are given. All computational testing is consistent with that shown. Computer runs were made on a VAX 11/780 with a VAX/VMS operating system and all floating-point quantities declared double precision.

The problem sets reported here were generated with a normal error distribution. A linear model of the form $X\beta = Y + \epsilon$ was assumed. True β values were generated from a uniform distribution over the interval $[0, 20]$ and X values from a uniform distribution over the interval $[-100, 100]$. From these values, the hyposthesized observed Y_i was set equal to $X_i\beta + \epsilon_i$, where ϵ_i was normal with a mean of zero and a variance of 400.

The initial phase of testing compared the primal and dual implementations. The first free observation was chosen as the observation to remove from the model at any stage. Both versions utilized an LU decomposition (Armstrong and Kung, 1982a; Murty, 1976) of the basis to solve linear equations, and any quantity with an absolute value less than 1.E-8 was taken to be zero. Table 1 presents the results with the primal algorithm (P) and the dual algorithm (D). The primal algorithm always required fewer updates, but because of the additional work in a primal iteration, the dual method was often faster. The major differences between the two algorithms were found when solving the larger problems, where the dual implementation was significantly faster. Since the solution of large problems in a reasonable amount of computer time is a primary objective, the primal algorithm was not considered in the future testing.

TABLE 1 Comparison of Primal and Dual Algorithms for the LAV Most I Influential Observations Problem[a]

	m = 5, K = 1		m = 5, K = 2		m = 10, K = 1		m = 10, K = 2	
	D	P	D	P	D	P	E	P
50	1.01	.86	29.55	29.39	5.62	4.16	113.8	99.5
100	(138)	(53)	(4370)	(2246)	(352)	(159)	(116)	(4031)
100	4.46	3.97	183.9	177.3	34.96	22.24	631.1	639.9
	(475)	(9176)	(21015)	(7612)	(2160)	(616)	(35963)	(16485)
150	5.00	8.05	234.0	587.1	36.84	28.76	1468	1740
	(418)	(244)	(22679)	(17587)	(1863)	(573)	(80005)	(31931)
200	14.8	17.6	550.0	1407.	73.56	74.53	2756	4271
	(1182)	(429)	(39306)	(32856)	(3205)	(1152)	(134634)	(63184)

[a] The top value in each cell gives CPU seconds and the lower value give LU decomposition updates.

TABLE 2 Comparison of Three Branching Rules with the Dual Algorithm[a]

	m = 5			m = 10			m = 15		
K	D	D-N	D-R	D	D-N	D-R	D	D-N	D-R
1	1.01 (138)	1.07 (138)	1.08 (135)	5.62 (352)	4.93 (312)	5.05 (304)	24.4 (773)	22.9 (744)	19.6 (609)
2	29.55 (4370)	28.81 (4206)	29.28 (4046)	113.8 (7116)	111.1 (6928)	104.3 (5216)	313.4 (9129)	286.4 (8277)	268.6 (7498)
3	496.3 (73504)	485.6 (70899)	545.1 (60672)	1626 (97081)	1458 (85287)	1465 (81212)	4949 (144974)	4421 (127743)	4167 (114079)

[a] The value of n is 50. The top value in each cell give CPU seconds and the lower gives LU decomposition updates.

TABLE 3 Comparison of Three Branching Rules with the Dual Algorithm

	m = 5, K = 1			m = 5, K = 1			m = 5, K = 1			m = 5, K = 1		
	D	D-N	D-R	D	D-N	D-R	D	D-N	D-R	D	D-N	D-R
100	4.46	4.36	3.88	183.9	179.4	180.1	34.96	34.78	33.53	631.1	619.7	619.0
	(475)	(471)	(386)	(210145)	(20503)	(17935)	(2160)	(2157)	(2031)	(35963)	(35236)	(33516)
150	5.00	4.98	5.65	234.0	239.0	294.1	36.84	36.88	32.71	1468	1483	1533
	(418)	(418)	(397)	(22679)	(22787)	(23138)	(1863)	(1863)	(1595)	(80005)	(79743)	(78762)
200	14.8	14.7	14.7	55.0	562.8	680.9	73.56	76.19	78.48	2756	2979	3063
	(1182)	(1182)	(1075)	(39306)	(39056)	(37780)	(3205)	(3205)	(31334)	(134634)	(134634)	(137151)

The next phase of testing compared three different branching rules for the dual algorithm. In addition to choosing the first free observation to remove from the model, the following two rules were considered:

D-N: The first free nonbasic observation was chosen to be removed from the model. If all nonbasics were fixed in or out of the model, then the first free basic was chosen. The removal of a basic observation (associated π_i) requires more computations than the removal of a nonbasic observation. The rationale for the implementation labeled D-N is to reduce the number of times that basic observations have to be removed.

D-R: The free observation with the largest absolute residual was chosen to be removed from the model. Iy may be anticipated that the observations with large absolute residuals are more likely to be influential. This rule, D-R, attempts to obtain a small objective value early in the branching process and, thus, should not require the optimal solution for as many subproblems.

The results of comparing the three branching rules are summarized in Tables 2 and 3. There is no rule that is clearly superior. The number of LU decomposition updates is consistently reduced by the largest absolute residual rule, but often the additional labor required to implement the rule results in an overall increase in solution time. Table 2 clearly indicates how solution times increase in an exponential fashion as K increases.

5. CONCLUSIONS

The use of optimization techniques in statistical analysis has become more frequent in recent years. Specialized solution procedures have been incorporated in computer routines that are efficient and easy to use. These computer routines lay the groundwork for empirical testing from which an insight into statistical properties is obtained. Identifying outlying observations is an important problem in regression analysis, and this chapter has considered the problem when using a least absolute value criterion. A fundamental algorithm was developed and computational results with four implementations were presented. Through the use of implicit enumeration and special-purpose routines, it has been demonstrated that problems of realistic size can be solved. It must be emphasized that any outlier identification procedure can be considered only as one element in the process of examination of a data set for an outlier, not as the sole determinant. We propose the procedure given here to be used as a tool in exploratory data analysis. A computer code written in standard FORTRAN and designed to assist in identifying outlying observations when using a least absolute value criterion is available from the authors.

REFERENCES

Armstrong, R. D., and Kung, M. T. (1982a). A dual algorithm to solve linear least absolute value approximation. J. Oper. Res. Soc. 33: 931-936.

Armstrong, R. D., and Kung, M. T. (1982b). An algorithm to select the best subset for a least absolute value problem. TIMS Stud. Manage. Sci. Optimization Stat. 19: 931-936.

Armstrong, R. D., Frome, E. L., and Kung, D. (1979). A revised simplex algorithm for the absolute deviation curve-fitting problem. Commun. Stat. Simul. Comput. B8: 175-190.

Barnett, V., and Lewis, T. (1980). Outliers in Statistical Data. Wiley, New York.

Barrodale, I., and Roberts, F. D. K. (1973). An improved algorithm for discrete L_1 linear approximation. SIAM J. Numer. Anal. 10 (October): 839-848.

Barrodale, I., and Roberts, F. D. K. (1974). Solution of an overdetermined system of equations in the L_1 norm. Commun. ACM 17 (June): 319-320.

Bartels, R. H., and Golub, G. H. (1969). The simplex method of linear programming using LU decomposition. Commun. ACM 12: 266-268.

Bassett, G., and Koenker, R. (1978). Asymptotic theory of least absolute error regression. J. Am. Stat. Assoc. 73: 618-622.

Charnes, A., Cooper, W. W., and Ferguson, R. (1955). Optimal estimation of executive compensation by linear programming. Manage. Sci. 2: 138-151.

Darboux, G., ed., Oeuvres De Fourier, vol. 2, pp. 324-325. Gauthier-Villars, Paris.

Dielman, T., and Pfafferberger, R. (1982). LAV (least absolute value) estimation in linear regression: A review. TIMS Stud. Manage. Sci. Optimization Stat. 19: 31-52.

Gentle, J. E. (1977). Least absolute value estimation: An introduction. Commun. Stat. Simul. Comput. B6 (4): 313-328.

Gentle, J., Kennedy, W., and Sposito, V. (1977). On least absolute value estimation. Commun. Stat. Theory Methods A6: 839-845.

Gentleman, J., and Wilk, M. (1975). Detecting outliers. II. Supplementing the district analysis of residuals. Biometrics 31 (June): 387-410.

Kennedy, W. J., and Gentle, J. E. (1980). Statistical Computing. Marcel Dekker, New York.

Koenker, R., and Basset, G., Jr. (1978). The asymptotic distribution of the least absolute error estimator. J. Am. Stat. Assoc. 73: 618-622.

Mickey, M. R. (1974). Detecting outliers with stepwise regression. Commun. UCLA Health Sci. Facility I: I.

Mickey, M., Dunn, O., and Clark, V. (1967). Note on the use of stepwise regression in detecting outliers. Comput. Biomed. Res. 1: 105-222.

Murty, K. (1976). Linear and Combinatorial Programming. Wiley, New York.

5

Fitting Redescending M-Estimators in Regression

STEPHAN MORGENTHALER* Yale University, New Haven, Connecticut

1. INTRODUCTION

Regression is not only a methodology for fitting preconceived equations to data which follow the model quite closely. It is also a tool for data description. In the context of description it can make sense for a regression program to give several distinct answers. Sometimes linear model data allows several fits, maybe with associated regions of use in the space of explanatory variables. Such multiplicity can occur in at least two different manners. First, different *subpopulations* might require different parameter values. Second, we might have data from a *nonlinear system* allowing different stable equilibria. In an observational study of such a system we may observe several of these equilibria, resulting in multiple regression surfaces (for examples, see Cobb and Zacks, 1985).

Regression diagnostics and bounded-influence regression both address this problem to a limited degree. Even though such procedures usually give a single answer, they try to avoid averaging distinct surfaces.

*Current affiliation: Ecole Polytechnique Federal de Lausanne, Lausanne, Switzerland.

Instead, they pick the solution which apparently fits the majority of the data. Both of these technologies concentrate mainly on leverage points which may overwhelm the fit. There is a large literature on the subject of leverage points and influence, and the interested reader should consult Belsley et al. (1980) or Cook and Weisberg (1982).

Another topic which is relevant to our discussion is the problem of high breakdown point. Suppose up to one-half of the data is garbage. Is there a method which is guaranteed to find the fit to the more than one-half good data points? This is not intended as a realistic model. But as a toy in inference it has proved to be a hard puzzle. The existing solutions are due to Siegel (1982) and Rousseeuw (1984). A regression estimator with a high breakdown point is the ideal first step in the search for multiple answers. It is guaranteed to find the most important surface, and one can imagine a sequential uncovering of further substructures through the deletion of observations.

This chapter discusses the application of <u>redescending M-estimators</u> to the problem of multiple solutions. The class of estimators considered here is related to the S-estimators proposed in Rousseeuw and Yohai (1984). They concentrate on the absolute minima of an implicitly defined criterion function and prove consistency and asymptotic normality for those solutions. They also show that a breakdown point close to 50% can be achieved with these estimators. We will not be concerned here with such statistical properties, but rather discuss computational issues. The main reason for this attitude is the lack of a plausible, widely applicable stochastic model for the kind of data we are interested in. If there appear to be three solutions in a data set with n = 20, for example, it is not at all clear that all of these—or even one—have any intrinsic importance if we really could let n tend to infinity. On the other hand, the description of the data with three solutions might be much more accurate. This discussion raises the important question of how to interpret the results of an analysis. But this must depend on the individual data set.

M-estimators are well known among statisticians, but the redescending type has been rejected as often as it has been commended. The main difficulty seems to be a computational one. To compute such an estimator, one appears to need a safe starting value such as a high breakdown estimator. There are, however, alternative computational methods which might turn out to be easier to implement and to calculate than any of the high breakdown estimators. We will end the introductory comments by briefly reviewing the use of a particular redescending M-estimator in the location case.

Despite its nonoptimality (or possibly because of it), Tukey's <u>biweight</u> (Beaton and Tukey, 1974; Tukey, 1977) has become popular. Let

$$\eta(x) = \begin{cases} -(1 - x^2)^3/6, & |x| \leqslant 1 \\ 0, & |x| > 1 \end{cases} \tag{1}$$

The biweight location estimator is defined as the point $T_n^k(y_1, \ldots, y_n)$ minimizing the criterion function

$$C_k(t) = \sum_{i=1}^{n} \eta\left(\frac{y_i - t}{k}\right) \quad (k > 0) \tag{2}$$

From (1) we see that $\eta(x)$ is, up to a normalizing constant, the negative of a smooth, symmetric, and unimodel probability density. The function $C_k(t)$ is hence the negative of a kernel density estimate with kernel width k. As k gets smaller, the corresponding density estimate becomes less smooth, resulting in a criterion function with a multitude of local minima. Freedman and Diaconis (1982) discuss the inconsistency of $T_n^k(y_1, \ldots, y_n)$, when the underlying distribution is multimodal and k is smallish. In their opinion, a consistent estimator always should estimate the center of a symmetric distribution. The biweight clearly does not do that.

The preferred algorithm for finding T_n^k, at least approximately, consists in choosing k based on the data $y_1, \ldots, y_n$. Then an iteration is performed with

$$t_{new} = \frac{\Sigma_{l=1}^{n} w_l(t_{old})y_l}{\Sigma_{l=1}^{n} w_l(t_{old})}$$

until convergence. The iteratively computed weights are

$$w_l(t) = \begin{cases} \dfrac{\eta'[(y_l - t)/k]}{(y_l - t)/k} = \left[1 - \left(\dfrac{y_l - t}{2}\right)^2\right]^2, & \text{if } \dfrac{|y_l - t|}{k < 1} \\ 0, & \text{otherwise} \end{cases}$$

This algorithm converges, but not necessarily to T_n^k, the global minima. The value to which the successive estimates converge depends on the starting estimate. Despite all these shortcomings, the estimator defined this way with starting value t_0 = median $(y_1, \ldots, y_n)$ and k = 6× (median absolute deviation) is an excellent estimator of location.

2. CRITERION FUNCTIONS FOR REGRESSION

2.1 The Construction of a General Criterion

Consider a situation where the data we have available are n pairs (y_1, x_1), ..., (y_n, x_n) with $y_i \in \mathbf{R}$ and $x_i \in \mathbf{R}^{p-1}$ $(i = 1, \ldots, n)$. We wish to fit structures of the form

$$y = \mu + \beta' x + \text{error}$$

where $\beta \in \mathbf{R}^{p-1}$ and $\mu \in \mathbf{R}$. As we mentioned in the introduction, several different values of the parameter $\theta = (\mu, \beta) \in \mathbf{R}^p$ might turn out to give a reasonable description of at least a sizable subset of the data. In a sense, we want to solve a problem in clustering, namely to identify hyperplanes of the form $y = \mu + \beta' x$ in $\mathbf{R}^p$ around which many of the observations cluster.

An alternative way of thinking involves the space of the θ values, that is, $\mathbf{R}^p$. For each observation (y_i, x_i) let us define a (p - 1)-dimensional affine subspace $H_i = \{\theta = (\mu, \beta) \mu + \beta' x_i = y_i\}$. In other words, H_i contains the values of the parameter θ that fit the ith observation perfectly.

Remark: The subspace H_i is not well defined if $x_i = 0$ and we enforce $\mu = 0$. This can happen only if we do not fit an intercept but have available an observation at the origin. Note that such observations are of no use to us for fitting purposes and might as well be disregarded.

In general configuration the intersection of a selection of p sets from H_1, ..., H_n will contain a single point. In special cases this intersection may be empty or it may be an affine set. The latter occurs if some of the explanatory variables have been observed repeatedly, or if there are other linear dependencies among the explanatory variables.

Definition 1. For a collection $\{H_i : i = 1, \ldots, n\}$ of (p-1)-dimensional affine sets $\subset \mathbf{R}^p$, denote by $S(H_1, \ldots, H_n)$ the set of all points of intersection. By this we mean a point $t \in \mathbf{R}^p$ with the property $\{t\} = H_{i_1} \cap \cdots \cap H_{i_p}$ for some $\{i_1, \ldots, i_p\} \subset \{1, \ldots, n\}$.

In general configuration the linear subspaces H_1, ..., H_n generate $\binom{n}{p}$ points of intersection.

The problem now consists in finding clusters of the points in $S(H_1, \ldots, H_n)$. The reasoning behind this objective is as follows. Each point of intersection corresponds to a value for the parameter θ which gives a perfect fit to p or more of the n observations. If these points of intersection cluster, we have identified a parameter value which fits many of the observations.

Remark: The straightforward approach of computing (at least a sample of) the points of intersection has been explored elsewhere (Siegel, 1982). It leads to an elegant estimator with breakdown point of 50%. The computation, however, is a problem of combinatorial magnitudes, and it is desirable to avoid the calculation of these points when n or, especially, p is large.

Figure 1 shows a hypothetical picture with a single, fairly well established cluster. With the help of this picture, we want to discuss

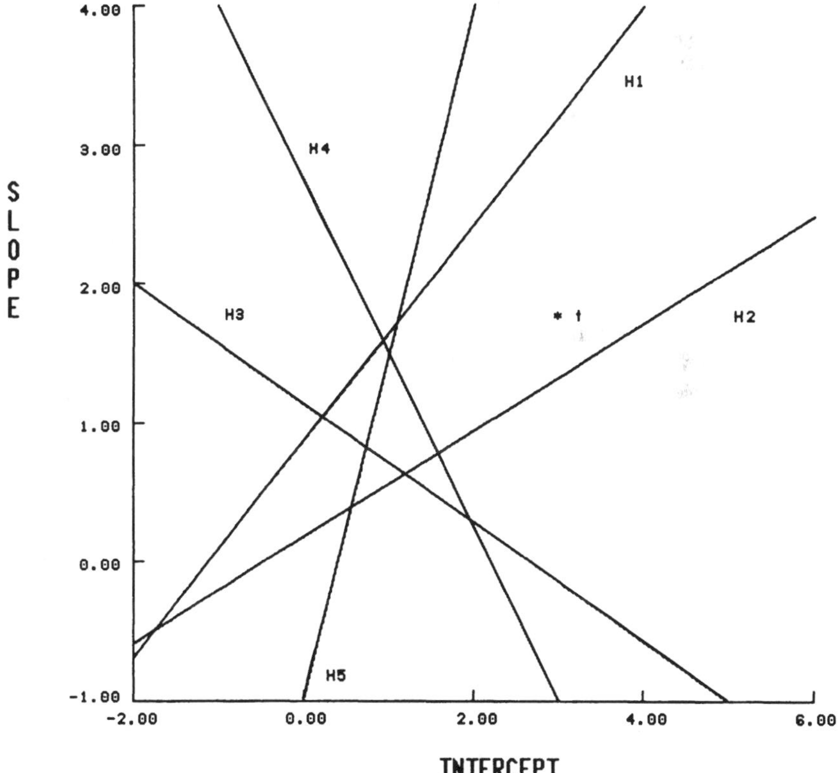

FIGURE 1 A regression data set with five observations. The line H_i contains the values of (intercept, slope) which fit the ith observation perfectly.

the construction of criterion functions, whose local minima are indicative for clusters.

Let $t \in \mathbb{R}^p$ and let $d_l(t)$ be some measure of distance between t and H_l. If many of the $d_l(t)$'s are smallish, there is evidence for a cluster. But, since several clusters may be present, we ought to disregard very large values of $d_l(t)$ in the presence of small such values.

A general class of criterion functions fulfilling these specifications is given by

$$C(t) = \text{Scale}[d_1(t), \ldots, d_n(t)] \tag{3}$$

where Scale $[\cdot]$ is a general location-invariant and scale-equivalent positive real function. Jaeckel (1972) considered this type of criterion in a paper conceiving the use of ranks in linear models. Many robust regression estimators proposed in the literature also have a criterion function of the form (3) with $d_l(t)$ replaced by $r_l(t) = y_l - (1, x_l')t$.

Example: Rousseeuw (1984) studies $R(t) = \text{median}[|r_l(t)|, l = 1, \ldots, n]$. He shows that the global minimizer of R(t) has a breakdown point close to 50%.

Intuitively, it is appealing to choose the parameter θ in such a way that the residual scale is minimal when compared with other possible residual configurations. The choice of $d_l(t)$, however, deserves a bit more comment.

For most statisticians, $d_l(t) = r_l(t)$ is a nearly automatic choice. But in recent years, alternative choices have been the object of study. It seems natural, for example, to use the Euclidean distance, that is,

$$|d_l(t)| = \min_{\theta \in H_l} ||t - \theta||$$

$$= \frac{|y_l - (1, x_l')t|}{||(1, x_l)||} = \frac{|r_l(t)|}{||(1, x_l||} \tag{4}$$

where $||(1, x_l)||$ denotes the Euclidean norm $(1 + \Sigma_{k=1}^{p-1} x_{lk}^2)^{1/2}$. This seems to be a rather surprising choice for $d_l(t)$ and is, in fact, not acceptable, since linear transformations of the explanatory variables would not be handled invariantly. In order to apply (4) we must first standardize the explanatory variables. Or, we must replace the Euclidean norm in the denominator by a norm which is invariant under linear transformations of the explanatory variables. After standardization, the distance (4) leads to nothing else but the family of robust estimators proposed by Hill and Ryan (see Hill, 1977). The distance measure (4) points in the direction of generalized M-estimators and bounded influence estimators. For us, this is somewhat of a side topic, and we will be content with the choice of $d_l(t) = r_l(t)$. For additional comments on methods related to bounded influence, the key references are Krasker and Welsch (1982) and Hampel et al. (1986).

The primary topic of this chapter is the use of redescending M-estimators. They fit into the general prescription (3) and basically lead to a choice for the scale function.

> Let $\eta(x)$ be a twice continuously differentiable function which is symmetric around 0, negative on the open interval (-1, 1), zero outside (-1, 1), and monotone increasing on the interval (0, 1). Also assume that $\eta'(x)/x$ is positive and monotone decreasing on (0, 1). (5)

Example. Tukey's biweight function is a polynomial of the lowest possible degree satisfying all these contraints [see (1)].

The use of any as described in (5) leads to the criterion

$$S_k(t) = \sum_{l=1}^{n} \eta\left(\frac{d_l(t)}{k}\right), \quad d_l(t) = r_l(t) \tag{6}$$

where η satisfies the conditions in (5). As we mentioned in the introduction, this class is related to the S-estimators introduced by Rousseeuw and Yohai (1984).

Not all resistant methods in regression are covered by (3). As we have already mentioned, diagnostics offers another route. Most of the diagnostic work is cast in the language of outliers and influence. One identifies subsets whose deletion substantially changes the fit. Again, the computational problem is of combinatorial proportions, since many subsets must be tried. Part of the research effort has gone into searching for influential subsets without having to try all possible subsets. Again, clustering techniques suggest themselves.

2.2 Some Properties of $S_k(t)$

In the remainder of this section, we want to discuss the behavior of a criterion function of the form (6). Such functions always have a multitude of weak local minima, if by this we mean any point t_0 such that $S_k(t) \geqslant S_k(t_0)$ for all t in some open neighborhood of t_0. In Figure 1, for example, all the points on the lines H_l which are sole contributors to S_k—i.e., all the other terms in S_k are = 0—define a local minimum in this weak sense. The minima of interest to us, however, are those which correspond to a bowl shape of the surface S_k.

Definition 2. A local minimum t_0 of $S_k(t)$ is called degenerate if there exists a direction $h \in \mathbf{R}^p$ such that $S_k(t_0 + \lambda h) = S_k(t_0)$ for all $\lambda \geqslant 0$.

All nondegenerate minima lie in a bounded subset of the parameter space. Define the set

$$A_l(k) = \{t \in \mathbf{R}^p \mid |r_l(t)| < k\} \subset \mathbf{R}^p$$

for $l = 1, \ldots, n$. This set contains the t values which have a nonzero contribution of the lth observation to $S_k(t)$. Of importance is the set

$$A = \bigcup_{L \in \mathfrak{L}} \left[\bigcap_{l \in L} A_l(k) \right] = \bigcup_{L \in \mathfrak{L}} B_L \tag{7}$$

where $\mathfrak{L}$ contains subsets $L \subset \{1, \ldots, n\}$ of size p which satisfy the condition that $(1, x_l')'_{l \in L} \in \mathbf{R}^p$ are linearly independent.

Remark: It may help the reader to go back to Figure 1. In that picture all 10 pairs of distinct indices correspond to linearly independent x_l's. The set $\mathfrak{L}$, therefore, contains all pairs. The subsets $A_l(k)$ are strips $\subseteq \mathbf{R}^2$ with central line H_l and a width depending on x_l and k. The set A in (7) is the union of all pairwise intersections of such strips.

Clearly, $A = \cup_{i=1}^{I} C_i$, where $(C_i)_{i=1}^{I}$ is a family of nonempty, disjoint, bounded, open sets. Simply take unions of those B_L, $B_{L'}$ in (7) which have nonempty intersections to define the C_i's. The various properties of these C_i's are then easy to check.

Lemma 1. The set A defined in (7) contains all nondegenerate local minima.

Proof: Take $t \in \mathbf{R}^p - A = \cap_{L \in \mathfrak{L}} [\cup_{l \in L} (\mathbf{R}^p - A_l(k))]$. In other words, at most p - 1 observations with linearly independent explanatory values contribute to $S_k(t)$. There follows the existence of a linear subspace of dimension $\geqslant 1$ such that S_k is constant if we move away from t in the directions defined by this linear subspace and stay inside $\mathbf{R}^p - A$. The point t, if it is a local minimum at all, must be degenerate.

We can strengthen Lemma 1 in the following way, which is particularly interesting if $2 \leq I$, that is, A is partitioned into several disjoint pieces.

Lemma 2. Each of the sets $C_1, \ldots, C_I$ contains at least one nondegenerate local minimum.

Proof: (I). Take the case $I = 1$. The set A is a connected, bounded, open subset $\subset \mathbf{R}^p$. Denote the boundary of A by ∂A and the closure of A by $\bar{A} = A \cup \partial A$. S_k is a continuous function and attains its global minimum over $\bar{A}$. The value of the global minimum $S_k(t_g)$ is less than $-p$, since A contains points of intersection where S_k has value $-p$. From the definition (7) of A it follows that $S_k(t) > -(p - 1)$ for all $t \in \partial A$. It follows that $t_g \in A$ and that t_g is nondegenerate.

(II). Suppose $2 \leq I$. Denote by $\mathcal{L}_i \subseteq \mathcal{L}$ those sets of p indices indices involved in the construction of C_i $(i = 1, \ldots, I)$, that is, $C_i = \cup_{L \in \mathcal{L}_i} \cap_{l \in L} A_l(k)$. It follows that $d_l(t) > k$, whenever $l \notin \cup_{L \in \mathcal{L}_i} L = L(i)$ and $t \in C_i$.

Suppose this were wrong for some $l \notin L(i)$ and some $t \in C_i$. We could then conclude that $A_l(k) \cap C_i \neq \emptyset$. In other words, there would exist at least one $M \in \mathcal{L}_i$ with $A_l(k) \cap (\cap_{m \in M} A_m(k)) \neq \emptyset$. In such a set we can always find $p - 1$ elements which together with i form a set $J \in \mathcal{L}$; that is, they have linearly independent explanatory variables. But clearly

$$\left(\bigcap_{j \in J} A_j(k) \right) \cap \left(\bigcap_{m \in M} A_m(k) \right) \neq \emptyset$$

which implies $J \in \mathcal{L}_i$. This contradicts $l \notin L(i)$.

For values $t \in C_l$, therefore,

$$S_k(t) = \sum_{l \in L(i)} \eta \left[\frac{d_l(t)}{k} \right] = S_k^*(t)$$

where $S_k^*(t)$ denotes the criterion function evaluated for the data set $((y_l, x_l))_{l \in L(i)}$. Now, we can use case (I).

Other interesting questions could be asked about the local minima of $S_k(t)$. Of special interest is the number of these local minima as a function of k. In a similar setting, Reeds (1985) provides an interesting discussion. In our case it is obvious that if k is small, there are up to $\binom{n}{p}$ local minima. That gives an upper bound.

3. ITERATIVELY REWEIGHTED LEAST SQUARES (IRLS)

The optimization algorithm we will use throughout is iteratively reweighted least squares (IRLS). This algorithm was proposed in Beaton and Tukey (1974) and has been discussed in Huber (1981, Section 7.8) and Osborne (1985, Section 5.4).

A step in the application of this algorithm to S_k [see (6)] consists of computing the weighted least squares estimator t_{new} with weight $w_l(t_{old})$ on the lth observations, where

$$w_l(t_{old}) = \begin{cases} 0, & \text{if } |r_l(t_{old})| > k \\ \dfrac{\eta'(|r_l(t_{old})|/k)}{|r_l(t_{old})|}, & \text{otherwise} \end{cases} \tag{8}$$

The two books mentioned discuss the convergence of this algorithm for the case where $\eta'(x)/x$ is positive and nonincreasing on $(0, \infty)$. We have to extend this result to the case (5), where η is constant outside a finite interval.

The following lemma deals with the range of weighted least squares estimators. It is of use to us in formulating a regularity condition for the convergence proof but is of interest in its own right. As in Section 2, we denote by H_l the (p - 1)-dimensional affine set of parameter values which fit the lth observation perfectly. In determining the achievable parameter values under a weighted least squares scheme, the points of intersection $S(H_1, \ldots, H_n)$ (Definition 1) are of interest. We will prove that any weighted least-squares estimator lies within the closed convex hull $\text{conv}(S(H_1, \ldots, H_n))$. This theorem is important also in the context of jackknifing and bootstrapping regression estimates and has quite an old history (e.g., see Wu, 1986). Our proof is geometric in nature and we hope this will be of interest.

Lemma 3. Let $H_l = \{\theta = (\mu, \beta) : \mu + \beta' x_l = y_l\}$ and assume that $S(H_1, \ldots, H_n)$ (see Definition 1) is nonempty. If $t^* \in (\mathbf{R}^p$ - closed convex hull of $S(H_1, \ldots, H_n))$, then there exists $t \neq t^*$ such that $|r_l(t)| \leq |r_l(t^*)|$ for $l = 1, \ldots, n$.

The proof of this lemma can be found in the Appendix.

Corollary 4. Any solution to a weighted least squares problem with full rank must lie inside the closed convex hull of $S(H_1, \ldots, H_n)$.

The IRLS algorithm consists of successive computations of weighted least squares estimates with diagonal weight matrices. Let A be the set (7). It is clear that the weight matrix (8) will be singular unless $t \in A$. For the IRLS algorithm to converge, we must have a condition which ensures $t_{old} \in A \longrightarrow t_{new} \in A$. We can easily state a regularity condition which ensures just that by using Corollary 4. Unfortunately, this condition is tedious to check in practice. Unless k is quite small, one probably does not have to worry about the implication $t_{old} \in A \longrightarrow t_{new} \in A$. It is even possible that this implication is true for any value of k.

To state our regularity condition, we need an additional technical concept. We say that a set of indices $I \subset \{1, \ldots, n\}$ is connected if $\cap_{i \in I} A_i$ (k) (see Section 2.2) is nonempty.

Lemma 5. Suppose k has a value which ensures that the closed convex hull of $S[(H_i)_{i \in I}]$ is contained in A for all connected index sets I and suppose η satisfies the conditions (5). If $t_{old} \in A$, then $t_{new} \in A$ and $S_k(t_{new}) < S_k(t_{old})$ unless S_k has a local minima at t_{old}.

Proof: $t_{old} \in A \longrightarrow t_{new} \in A$ follows from Corollary 4.

The monotonicity of the algorithm follows from the idea described in Dutter (1975). Consider the quadratic function

$$u_l(t) = \eta\left[\frac{|r_l(t_{old})|}{k}\right] + \frac{w_l(t_{old})}{2k}\left[r_l(t)^2 - r_l(t_{old})^2\right] \quad (w_l(t_{old}) > 0)$$

where $w_l(t_{old})$ is the weight from (8) and $r_l(t) = y_l - (1, x_l')t$. For observations with $w_l(t_{old}) = 0$, put $u_l \equiv 0$. It is now straightforward to check that $U(t) = \Sigma\, u_l(t)$ is minimized at t_{new}. Furthermore, $U(t_{old}) = S_k(t_{old})$ and $S_k \leqslant U$. Therefore, we conclude that

$$S_k(t_{new}) \leqslant U(t_{new}) < U(t_{old}) = S_k(t_{old})$$

unless U is already minimized at t_{old}.

Under the conditions of Lemma 5, the successive iterates t_0, t_1, ... stay inside the bounded set A. Consequently, this sequence has points of accumulation. Each of those is a nondegenerate local minimum of S_k.

Proposition 6. Under the conditions of Lemma 5, the IRLS converges for any starting value $t_0 \varepsilon A$.

The criterion function S_k [see (6)] generally has more than one non-degenerate local minimum. It is, therefore, insufficient to run the IRLS algorithm only once. Instead, one needs strategies for finding a set of plausible starting values. This is the topic of the next section.

4. ALGORITHMS FOR FINDING MULTIPLE MINIMA

Let $(x_1, y_n), \ldots, (x_n, y_n)$ denote our data and define $S_k(t)$ as in (6), where we postulate the conditions (5) for η. We then know that $S_k(t)$ has a continuous second derivative for each $k > 0$. The extremal points of interest to us are those $t_0 \varepsilon \mathbf{R}^p$ with gradient $[S_k(t_0)] = 0$ and no degeneracy. Generally, $S_k(t)$ will have several such local minima, and the topic of this section is the enumeration of these extremal points. We recommend the use of the biweight criterion [see (1)]. For the biweight we can then say something about the parameter k. A commonly used value is $k = 6\text{MAD}(\hat{\theta}_{LS})$, where

$$\text{MAD}(t) = \text{median absolute deviation of } \{r_1(t), \ldots, r_n(t)\}$$

and $\hat{\theta}_{LS}$ denotes the least squares estimate of θ. It is obvious that very small values of k are of no help to us. The criterion $S_k(t)$ generally has $\binom{n}{p}$ global minima when k is small enough, one for each point of intersection (see Section 2).

Small values of k show us too much detail. If, on the other hand, k is very large, then $S_k(t)$ resembles the least squares criterion and not enough detail is shown for our purposes. Should there really be two solutions, we would settle on some average of the two with such a criterion. The emphasis in the following subsections is on procedures which can be implemented in any flexible statistical package with macro capability.

4.1 Complete Search

The only method guaranteed to find all the local minima of $S_k(t)$ is a search among all $t \varepsilon \mathbf{R}^p$. Suppose we want to search a p-dimensional box $(a_1, b_1) \times (a_2, b_2) \times \cdots \times (a_p, b_p)$. Then for any integer m we could compute $S_k(t)$ on the points of a grid with grid size $\Delta = (b_i - a_i)/m$ in

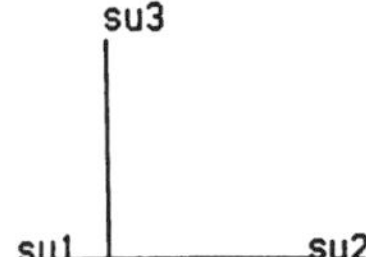

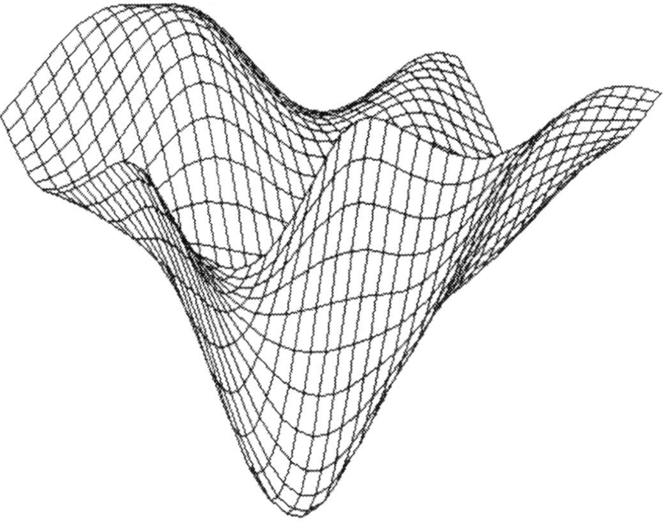

FIGURE 2 The surface shown represents the criterion function $S_k(t)$ [see (6)] based on the biweight. There is apparently a single global minimum. The tripod in the upper right-hand corner shows the direction of the axis for intercept (su1), slope (su2), and criterion (su3).

ith component of t. This would mean $(m + 1)^p$ evaluations of $S_k(t)$. Figure 2 shows such a grid for the data already used in Figure 1.

Retaining all grid points with values of S_k less than the values of its 2p neighbors will lead to a finite set of starting values $T \subset \mathbf{R}^p$. Using all these starting values with the IRLS algorithm, we will find all the local minima within the searched box as long as the number m is chosen large enough. The choice of the box to be searched can be quite a difficult one to make in practice. Furthermore, a complete search is realistic only for small values of p, because of the computational complexity.

4.2 Using the Residuals from the Previous Fit

It is desirable to find methods which, while not guaranteed to locate all minima, will do so under favorable circumstances. A much simpler idea than a complete search would be a sequential search. Suppose we found one local minimum. Can we use the residuals from this particular fit to find the next local minimum? If the criterion function $S_k(t)$ has only two local minima, i.e., if two reasonably well defined regression surfaces exist, the following idea might work. Once we found one of the solutions, we would hope that the corresponding residuals cluster into one set distributed around zero and a second set of large residuals. Using the observations with large residuals, we can then compute a starting value t_0 from which IRLS will find the second solution.

A formalization of this idea is as follows. Use the starting value $\hat{\theta}_{LS}$ and apply IRLS until convergence. Denote this solution by $\hat{\theta}_1$. The final weights from the iteration are then

$$w_l = \frac{\eta'(r_l(\hat{\theta}_1))}{r_l(\hat{\theta}_1)} \quad \text{if } r_l(t) \leqslant k \text{ and } w_l = 0 \text{ otherwise} \tag{9}$$

or, in the biweight case,

$$w_l = [k^2 - r_l(\hat{\theta}_1)^2]^2 \quad \text{if} \quad r_l(\hat{\theta}_1)^2 \leqslant k \quad \text{and} \quad w_l = 0 \text{ otherwise}$$

Next, compute the weighted least squares estimator $\hat{\theta}_{wls}(v)$ with weights $v_l = \text{inv}(w_l)$, where inv(·) is a function which inverts the size of weights. Two examples of inv(·) are

$$\text{inv}(w) = \frac{1}{\delta + w}$$

$$\text{inv}(w) = 1 - w + \delta \tag{10}$$

where δ is a small number such as $\delta = 0.01$. Use $\hat{\theta}_{wls}(v)$ as the starting value for a second round of the iteratively reweighted least squares algorithm to find $\hat{\theta}_2$.

This procedure will work well in balanced designs (see Section 6 for an example). If there are leverage points, or if the design is otherwise unbalanced, the hope that the residuals will cluster if there are two or more solutions may be optimistic.

It is possible to iterate the idea of inverted weights. Each solution already found will have an associated n-vector of final weights. Let us denote these by $\mathbf{w}_1$, $\mathbf{w}_2$, etc. If we want to find the next solution, we have to use the points which do not yet have a good fit. Inverting the weight vector $\mathbf{w} = \max\{\mathbf{w}_1, \mathbf{w}_2, \ldots\}$ and using inv($\mathbf{w}$) to compute a starting guess will do exactly that (J. W. Tukey, personal communication).

From the huge literature on regression for unbalanced designs it is clear that special precautions are necessary in such cases. The next subsection discusses a simple approach to this problem.

4.3 Making Use of the Design

One way in which two regression surfaces can appear is through peculiarities in a design, either if the true surface is curved or if the error distribution has heavy tails. Even though the data are unable to tell us exactly what the underlying reason is, they do indicate that something is wrong by admitting several solutions. If we investigate the possibility of a nonplanar surface, the region of the space of explanatory variables where the surface is to be used becomes important. (A classical paper on this subject, although in a different context, is Box and Draper, 1959.) Under such considerations, the design points should span the region of interest roughly uniformly. If the design grossly deviates from this uniform pattern, a single region of interest might be inappropriate. Statistical techniques which deal with this question are built on an analysis of the covariance structure of the explanatory variables.

We wish to detect whether the design points show distinct clusters in the space of explanatory variables. A helpful graphical device is to plot the eigenvectors associated with the two largest eigenvalues of the design covariance matrix against each other. If distinct clusters are present, we might want to choose an increasing sequence $R_1 \subseteq R_2 \subseteq \ldots \subseteq R_m$ of regions including more and more of the observations. Typically $m = 2$ and R_2 captures a few outliers together with the great majority in R_1.

Fitting inside each of these regions separately will lead to m different fits. Some of these might differ quite drastically. One interpretation is clear. Enlarging the region of interest changes the surface, because the true surface is nonplanar. As we already emphasized, the ultimate cure to this problem might not be found in the data alone. Instead, more measurements or expert judgment might be required.

The final step would again be the application of the iteratively reweighted least squares algorithm. The m solutions corresponding to the regions $R_1, \ldots, R_m$ would be used in turn as starting values.

The treatment of "outliers" in the space of explanatory variables along these lines seems more appropriate than the methods using generalized M-estimators, because the choices involved are brought out in the open.

4.4 Treating k as a Free Parameter

Finally, we want to mention numerical methods designed to find global minima of criteria like $S_k(t)$ which might have several extremal points. One possibility is to use k as a free parameter. As k is lowered from a large value toward zero, the surface $S_k(t)$ changes from a bowl shape to a shape with very many global minima. Now, apply iteratively reweighted least squares for each k value of a finite decreasing sequence $k_0, k_1, k_2, \ldots$. As a starting value for each iteration use the value to which the previous iteration converged.

This type of procedure will find the global minima of $S_k(t)$ as long as k is lowered slowly enough. But the slower k is decreased, the more intensive the computations. We found that steps for k on the order of $\mathrm{MAD}(\hat{\theta}_{LS})$ seem to work at least sometimes.

Treating k as a free object has the additional advantage that we do not need to specify a lower boundary of the sort $6 \times \mathrm{MAD}(\hat{\theta}_{LS})$. Instead, we may simply keep lowering k as long as the total weight $w_1 + \cdots + w_n$ [see (8)] is bigger than a prechosen fraction of n, for example, 3n/4.

The methods discussed in Sections 4.2 and 4.3 can be used in conjunction with the algorithm we described. But this does not seem to add much. The algorithm we have discussed here is useful when we want to find the global minima without having to enumerate all extremal points.

5. EXAMPLES

In this section we use two data sets to demonstrate the techniques for fitting a biweight regression.

5.1 Agriculture in Eleven Countries

This data set is compiled from the U.N. statistical yearbook. The raw data are given in Table 1 and plotted on a logarithmic scale in Figure 3. It is immediately clear that two lines fit the data substantially better than a single line could. This is a case where an essential variable is missing. Note, however, that both slope and intercept change when we go from one solution to the other, indicating interaction terms between the given variables and the missing variable.

Since p = 2, a complete search is rather simple, and Figure 4 shows that such a strategy would be successful. The two distinct solutions are clearly visible. For the tuning constant, the value 3.3 = 6 × the median absolute deviation (MAD) of the least squares residuals is used throughout.

TABLE 1 Number of Eggs (Millions) and Number of Pigs (Thousands) Produced in 11 Countries During 1970

Country	Eggs	Pigs
Angola	86	170
Cameroon	176	92
Egypt	1,377	268
Kenya	432	153
Senegal	127	149
Czechoslovakia	3,733	6,310
Denmark	1,485	5,924
France	12,200	3,794
Germany	15,377	5,112
Hungary	3,280	3,174
Switzerland	693	3,353

In this example the design variable log(eggs) is well behaved, and we do not need to consider splitting the region of fit. What remains to be considered, then, is a sequential search with inverted weights. Using the least squares solution as a starting point, the IRLS with weights (8) quickly converges to the line L1 (see Figure 3). The final weights are

$$\mathbf{w}_1 = 0.97 \quad 1.00 \quad 0 \quad 0 \quad 0.76 \quad 1.00 \quad 0.77 \quad 0.96 \quad 1.00 \quad 1.00 \quad 1.00$$

Starting from the weighted least squares solution with inverted weights—using either of the functions in (10)—leads to the line L2. Now the final weights are

$$\mathbf{w}_2 = 1.00 \quad 0.78 \quad 1.00 \quad 1.00 \quad 0.91 \quad 0 \quad 0 \quad 0 \quad 0 \quad 0 \quad 0$$

The maximum $\max(\mathbf{w}_1, \mathbf{w}_2)$ equals

$$1.00 \quad 1.00 \quad 1.00 \quad 0.91 \quad 1.00 \quad 0.77 \quad 0.96 \quad 1.00 \quad 1.00 \quad 0.99$$

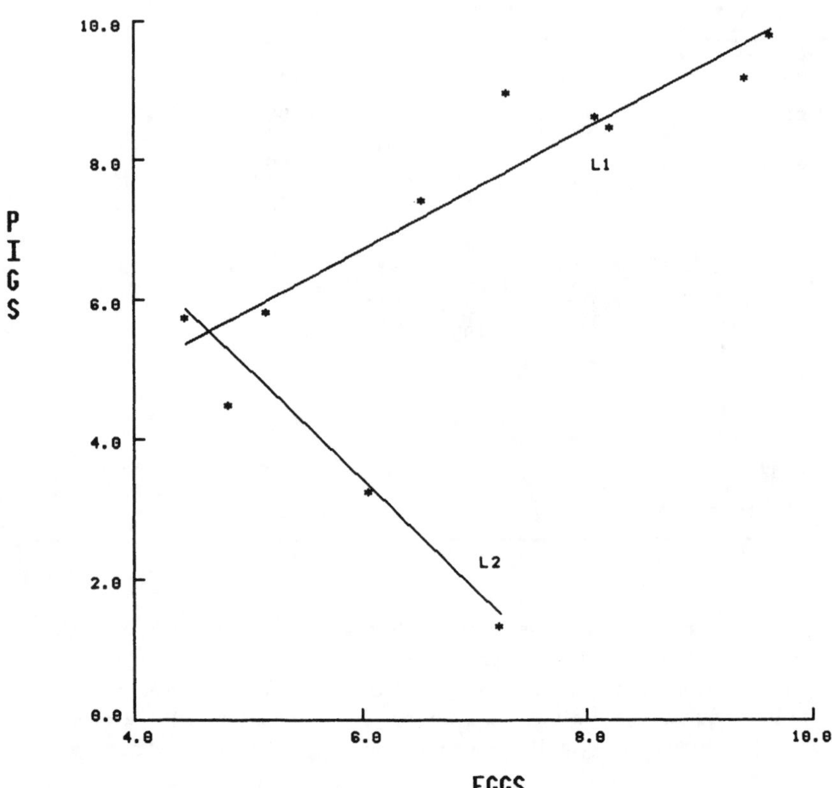

FIGURE 3 The number of pigs (in thousands) is plotted against the number of eggs (in millions). These values refer to the production during 1970. Eleven countries have been selected for this data set. Also included are the two regression lines, L1 and L2, which are uncovered through the use of the biweight criterion.

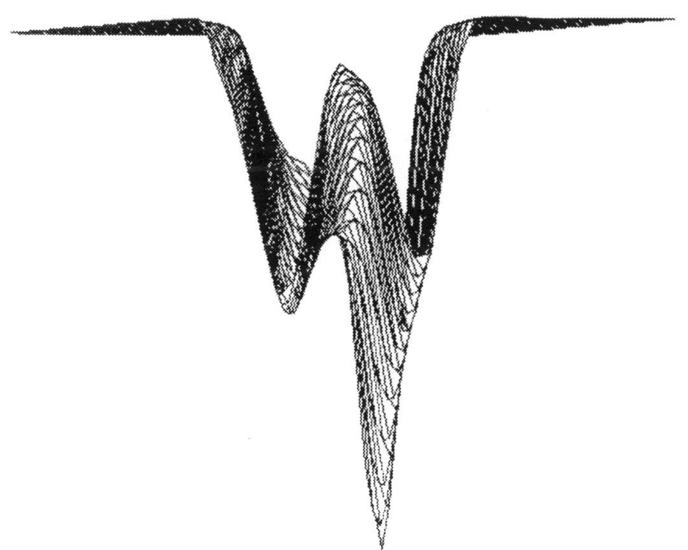

FIGURE 4 The plot shows $S_k(t)$ [see (6)] based on the biweight. The data used are the production values from Table 1. Two separated minima can be perceived. The tripod in the upper right-hand corner shows the direction of the axis for intercept (su1), slope (su2), and criterion (su3).

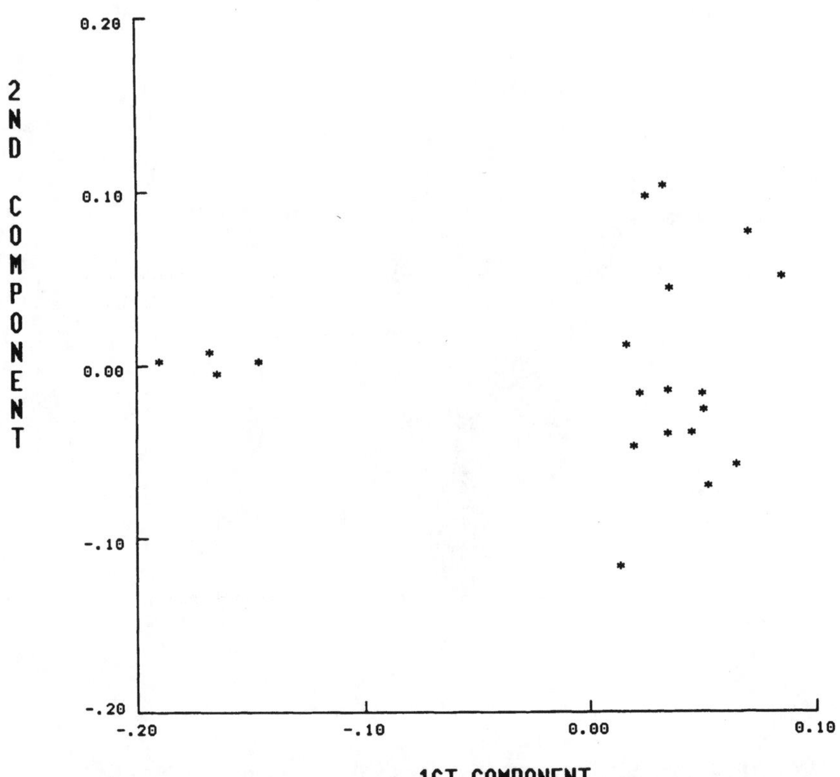

FIGURE 5 A plot of the first principal component against the second principal component of Rousseeuw's artificial data set. Four points are clearly separate from the others.

This weight vector leads back to the line L1 when we invert and calculate a new starting value.

5.2 Wood Specific Gravity with Artificial Outliers

The second data set is from Rousseeuw (1984, Table 2). It is an artificial construct based on an example from Draper and Smith (1981, p. 404). A look at the plot of the first two principal components (Figure 5) shows two distinct clusters. In such situations the method discussed in Section 4.3 makes a difference and is hence relevant. In this particular example, the artificial outliers occur exactly within the smaller cluster. Using only the bigger cluster of data values, or using all the data values, leads to two quite distinct solutions.

We ought to be careful in interpreting these situations. Figure 5 makes it apparent that the reduced data set spans a much smaller region in the space of explanatory variables than does the full data set. The equation derived from the reduced data set should be used only inside the reduced region. At the four outlying points, a serious bias could result from the use of the equation derived without those points.

For both data sets the technique discussed in Section 4.4 leads to the absolute minima down to very small values of k.

APPENDIX: Proof of Lemma 3

The proof of Lemma 3 presented here is geometric in nature. The conclusions we will state during the proof are not reduced to simple facts of set theory and logic. Geometric intuition is required.

Take $t^* \in \{\mathbb{R}^p - \text{closed convex hull of } S(H_1, \ldots, H_n)\}$. We want to show that the set

$$\bigcap_{l=1}^{n} \{t : |r_l(t)| \leq |r_l(t^*)|\} - \{t^*\}$$

is nonempty.

Each of the sets $\{t : |r_l(t)| \leq |r_l(t^*)|\}$ is a "sandwich" defined by hyperplanes parallel to H_l, one of them passing through t^*. To prove the lemma we need to consider only the immediate vicinity of t^*. Let H_l^* denote the hyperplane parallel to H_l passing through t^*, and let $\mathcal{G}_l$ denote the half-space bordered by H_l^* containing H_l. Another way of defining this half-space is $\mathcal{G}_l = \{t : r_l(t^*) \leq r_l(t) \text{ if } r_l(t^*) < 0 \text{ or } r_l(t) \leq r_l(t^*) \text{ if } r_l(t^*) > 0\}$. A special definition is required in the case where $r_l(t^*) = 0$. In that case we put $\mathcal{G}_l = H_l = H_l^*$.

Let $\mathcal{D}$ denote the intersection $\mathcal{D} = \cap_{l=1}^{n} \mathcal{G}_l$. Obviously, $t^* \in \mathcal{D}$. We show next that if $t^* \varepsilon \{\mathbf{R}^p$ - closed convex hull of $S(H_1, \ldots, H_n)\}$, then $\mathcal{D} \neq \{t^*\}$.

Let us first look at the special case where the design points are such that any p different H_l's intersect in a point. As a representative for this special problem we may take the case $p = 2$ with $x_i \neq x_j$ when $i \neq j$. (It will help the reader to consult Figure 1 and maybe to draw the lines H_l^*.) The point t^* then partitions the line H_l^* into two half-lines, either of which is denoted by $\widetilde{H}_l^*$. For any $m \geq 2$ the intersection $\mathcal{G}_1 \cap \cdots \cap \mathcal{G}_m$ is either

1. $\{t^*\}$ (0-dimensional) or
2. a half-line $\widetilde{H}_{l1}^*$ emanating from t^* (1-dimensional) or
3. a set with straight-line boundaries $\widetilde{H}_{l1}^*$ and $\widetilde{H}_{l2}^*$ and vertex t^* (2-dimensional).

That $\mathcal{G}_1 \neq \{t^*\}$ is trivial. Suppose $\mathcal{G}_1 \cap \cdots \cap \mathcal{G}_m \neq \{t^*\}$ for $m \geq 2$ and hence either from case 2 or case 3. Now if $\mathcal{G}_1 \cap \cdots \cap \mathcal{G}_m \cap \mathcal{G}_{m+1} = \{t^*\}$, then we have a contradiction to the assumption that t^* is outside the closed convex hull of $S(H_1, \ldots, H_n)$. In case 3, the lines H_{m+1}, H_{l1}, and H_{l2} form a triangle containing t^*. In case 2, take the lines H_{m+1} and H_{l1} together with any of the others among $H_1, \ldots, H_m$ to reach the same conclusion. For higher dimensions ($p \geq 2$) the same proof works, except that the number of cases to be considered grows to $p + 1$. The intersection $\mathcal{G}_1 \cap \cdots \cap \mathcal{G}_m$ ($m \geq 2$) can now have dimension $d = 0, 1, \ldots,$ or p. If we assume that $\dim(\mathcal{G}_1 \cap \cdots \cap \mathcal{G}_m) = d$ (≥ 1) but that $\mathcal{G}_1 \cap \cdots \mathcal{G}_m \cap \mathcal{G}_{m+1} = \{t^*\}$, we arrive at a contradiction as before. There must be $p - d$ residuals among $r_1(t^*), \ldots, r_n(t^*)$ which are equal to zero and d observations (not necessarily unique) which define the boundary of $\mathcal{G}_1 \cap \cdots \cap \mathcal{G}_m$. The corresponding p hyperplanes from $H_1, \ldots, H_m$ together with H_{m+1} form a closed polyhedron containing the point t^*.

In general, we cannot assume that any p H_l's will intersect in a point and the proof outlined above does not go through in its last step. Furthermore, a reduction of dimensionality of $\mathcal{G}_1 \cap \cdots \cap \mathcal{G}_m$ is now possible, not only because of residuals which are equal to zero but also because two hyperplanes may be parallel with t^* sandwiched between them.

Suppose $\dim(\mathcal{G}_1 \cap \cdots \cap \mathcal{G}_m) = d \geq 1$ but that $\mathcal{G}_1 \cap \cdots \mathcal{G}_m \cap \mathcal{G}_{m+1} = \{t^*\}$. We can find $0 \leq u \leq p - d$ residuals which are equal to zero and $0 \leq v = p - d - u$ cases of parallel hyperplanes among the first m observations. We denote the corresponding indices by $i1, \ldots, iu$ and $(j1, k1), \ldots, (jv, kv)$, respectively. The $u + v = p - d$ explanatory variables $(1, x_{i1}), \ldots, (1, x_{iu}), (1, x_{j1}), \ldots, (1, x_{jv})$ are linearly independent, since otherwise the dimension of $\mathcal{G}_1 \cap \cdots \cap \mathcal{G}_m$ would exceed d. Let us denote by l1, ..., ld the indices of the cases making up the boundary of $\mathcal{G}_1 \cap \cdots \cap \mathcal{G}_m$.

By considering the hyperplanes $H_{i1}, \ldots, H_{iu}, H_{j1}, \ldots, H_{jv}, H_{k1}, \ldots, H_{kv}, H_{l1}, \ldots, H_{ld}$ together with H_{m+1}, the contradictory fact that t lies inside the closed convex hull of $S(H_1, \ldots, H_n)$ follows. These hyperplanes intersect in a d-dimensional polyhedron which contains t^*.

What remains to be shown is

> if $\mathcal{D} \neq \{t^*\}$, then there exists $t \neq t^*$ which simultaneously reduces the size of all the residuals.

The set $\mathcal{D} = \mathcal{G}_1 \cap \cdots \cap \mathcal{G}_n$ has dimension ≥ 1. There exist, therefore, directions in which the residuals are simultaneously moved toward zero when we move away from t^* in one of those directions. If we move too far, some residuals will pass zero and start to grow in size. But in the immediate vicinity of t^* this cannot happen. In fact, there exists an $\epsilon > 0$ such that

$$\mathcal{D} \cap \{||t - t^*|| \leq \epsilon\} \subset \bigcap_{l=1}^{n} \{t : |r_l(t)| \leq |r_l(t^*)|\}$$

The value of this ϵ can be chosen as the minimum of the values of a corresponding ϵ appropriate for each case with nonzero residual.

This concludes the proof of the lemma.

ACKNOWLEDGMENT

The work on this paper has been partially supported by a grant from the Army Research Office (Grant No. DAAG29-84-K-0207).

REFERENCES

Beaton, A. E., and Tukey, J. W. (1974). The fitting of power series, measuring polynomials, illustrated on band-spectrographic data. Technometrics 16: 147-185.

Box, G. E. P., and Draper, N. R. (1959). A basis for the selection of a response surface design. J. Am. Stat. Assoc. 54: 622-654.

Cobb, L., and Zacks, S. (1985). Application of catastrophy theory for statistical modeling in the biosciences. J. Am. Stat. Assoc. 80: 793-802.

Cook, R. D., and Weisberg, S. (1982). Residuals and Influence in Regression. Chapman & Hall, London.

Draper, N. R., and Smith, H. (1981). Applied Regression Analysis. Wiley, New York.

Dutter, R. (1975). Robust regression: Different approaches to numerical solutions and algorithms. Res. Rep. No. 6. Seminar fur Statistik ETH, Zurich, Switzerland.

Freedman, D. A., and Diaconis, P. (1982). On inconsistent M-estimators. Ann. Stat. 10: 454-461.

Hampel, F. R., Ronchetti, E. M., Rousseeuw, P. J., and Stahel, W. A. (1986). Robust Statistics: The Approach Based on Influence Functions. Wiley, New York.

Hill, R. W. (1977). Robust regression when there are outliers in the carriers. Ph.D. thesis, Harvard University, Cambridge, Mass.

Huber, P. J. (1981). Robust Statistics. Wiley, New York.

Jaeckel, L. A. (1972). Estimating regression coefficients by minimizing the dispersion of residuals. Ann. Math. Stat. 43: 1449-1458.

Krasker, W. S., and Welsch, R. E. (1982). Efficient bounded-influence regression estimation. J. Am. Stat. Assoc. 77: 595-604.

Mosteller, F., and Tukey, J. W. (1977). Data Anslysis and Regression. Addison-Wesley, Reading, Mass.

Osborne, M. R. (1985). Finite Algorithms in Optimization and Data Analysis. Wiley, New York.

Reeds, J. A. (1985). Asymptotic number of roots of Cauchy location likelihood equations. Ann. Stat. 13: 775-784.

Rousseeuw, P. (1984). Least median of squares regression. J. Am. Stat. Assoc. 79: 871-880.

Rousseeuw, P., and Yohai, V. (1984). Robust regression by means of S-estimators. Robust and Nonlinear Time Series Analysis, J. Franke, W. Hardle, and D. Martin, eds. Lecture Notes in Statistics, No. 26. Springer, New York.

Siegel, A. F. (1982). Robust regression using repeated medians. Biometrika 69: 242-244.

Wu, C. F. J. (1986). Jackknife, bootstrap and other resampling methods in regression analysis. Ann. Stat. 14: 1261-1295.

6

On the Robustness of the Simple Linear Minimum Sum of Absolute Errors Regression

SUBHASH C. NARULA Virginia Commonwealth University, Richmond, Virginia

JOHN F. WELLINGTON Gannon University, Erie, Pennsylvania

1. INTRODUCTION

Let Y_i denote the value of a response variable corresponding to X_i, the value of a regressor (or predictor) variable for the ith observation, i = 1, 2, ..., n, where n is the number of observations. The simple linear regression model may be written as

$$Y_i = \beta_0 + \beta_1 X_i + \epsilon_i, \quad i = 1, 2, \ldots, n \tag{1}$$

where β_0 and β_1 are the unknown intercept and slope parameters, respectively, of the model, and ϵ_i represents the unobservable random error.

For the simple linear regression model (1), the least squares regression passes through $(\bar{X}, \bar{Y})$, the sample means of the regressor and the response variables, respectively. The least squares estimators of β_0 and β_1 are given by

$$\hat{\beta}_0 = \bar{Y} - \hat{\beta}_1 \bar{X} \tag{2a}$$

$$\hat{\beta}_1 = \frac{\Sigma_{i=1}^{n} (X_i - \bar{X})(Y_i - \bar{Y})}{\Sigma_{i=1}^{n} (X_i - \bar{X})^2} \tag{2b}$$

Clearly, $\hat{\beta}_0$ and $\hat{\beta}_1$ are determined and thus influenced by all the observations in the sample. These estimators are sensitive to outliers which are difficult to identify in regression models. In an excellent paper, Beckman and Cook (1983) discuss the diagnostic techniques that identify the influential observations and the robust regression procedures that accommodate the possibility of outliers in the least squares regression.

The minimum sum of absolute errors (MSAE) regression is considered a robust alternative to least squares regression by a number of authors. For example, Huber (1974, p. 927) stated that with regard to the L_p-estimator in regression, "p = 1 (MSAE regression) gives robustness in a technical sense (Hampel, 1971), i.e., resistance against arbitrary outliers." The MSAE regression is supposed to be less sensitive to outliers than least squares regression (Barrodale, 1968; Rice and White, 1964).

It has been shown by Karst (1958) and by Sposito et al. (1978) that for model (1), the MSAE regression passes through at least two observations; that is, the observed residuals (or errors) $e_i = Y_i - \tilde{Y}_i$ will be equal to zero for at least two observations, where $\tilde{Y}_i$ is the predicted value of the response variable for the ith observation. We shall call the observations with e_i equal to zero the defining observations; and the others, the nondefining observations. If the MSAE regression line passes through exactly two observations, it is completely determined by them. As such, it is not affected by the values of the other observations.

In fact, it is useful to note that the MSAE regression is to the least squares regression what the sample median is to the sample mean. For example, both the sample mean and the least squares estimators are determined and influenced by all the observations, whereas the sample median and the MSAE estimator are influenced by all the observations but determined by only a subset of observations. It is well known that the value of a sample median is unaffected if the magnitude of an observation is changed such that it remains on the same side of (either above or below) the sample median. A similar result is true for the MSAE regression. That is, the MSAE estimates of β_0 and β_1 are not altered by changes in the values of the response variable associated with the nondefining observations as long as these observations remain on the same side of (either above or below) the MSAE regression line. Since the MSAE regression is less susceptible to certain types of outliers, it is considered a robust alternative to the least squares regression.

In the analysis of the stack loss data (Brownlee, 1965, p. 454) using MSAE regression, Narula and Wellington (1985, pp. 184-185) found that the deletion of observation 18 had a negligible effect on the analysis, whereas the deletion of observation 17 had the maximum effect on the estimates of all the regression coefficients. This was surprising in light of the fact that observations 17 and 18 differed only in the value of x_3—the concentration of nitric acid in the absorbing liquid—but deletion of one observation had only a slight effect on the analysis whereas deletion of the other had the maximum effect! Further analyses of the data revealed that the estimators were affected only slightly as long as x_3 was greater than 78.073 but had a dramatic effect if x_3 was less than or equal to 78.073. Since the values of the variable x_3 for observations 17 and 18 lie on two sides of 78.073, this explains the divergent behavior of the two observations. Thus, from practical considerations, it will be very useful to know the intervals within which the value of the response variable and the value of the regressor variable can lie without changing the MSAE regression.

Our objective here is to develop algorithms to determine intervals for the value of the response variable and the value of the regressor variable (in the simple linear regression model) for the nondefining observations, ceteris paribus, within which the MSAE regression does not change. We also develop procedures to find intervals for the value of the response variable and the value of the regressor variable for the defining observations, ceteris paribus, within which the defining observations continue to be defining observations. Note that in the latter case, the MSAE estimates of β_0 and β_1 do change.

The rest of the paper is organized as follows: In Section 2 we develop algorithms to determine intervals on the value of the response variable and the regressor variable within which either the MSAE regression does not change or the defining observations remain the same. In Section 3 we illustrate the algorithms with an example. We conclude with a few remarks in Section 4.

2. METHODOLOGY

For the simple linear regression model (1), the MSAE regression passes through at least two observations. However, in the rest of the chapter, we assume that the MSAE regression passes through exactly two points. Suppose that it passes through observations (X_p, Y_p) and (X_q, Y_q). Then the MSAE estimates b_0 and b_1 of β_0 and β_1 are given by

$$b_0 = Y_p - b_1X_p = Y_q - b_1X_q \tag{3a}$$

and

$$b_1 = \frac{Y_p - Y_q}{X_p - X_q} \tag{3b}$$

Although a number of algorithms have been developed to determine the MSAE regression for model (1) (see Narula, 1982, 1987; Narula and Wellington, 1982; Dielman and Pfaffenberger, 1982, 1984), here we shall use the basic ideas of the algorithm due to Karst (1958). Without going into the details of finding (X_p, Y_p) and (X_q, Y_q) through which the MSAE regression passes, Karst's algorithm requires the following computations.

Define $Z_i(t) = Y_i(t)/X_i(t) = (Y_i - Y_t)/(X_i - X_t)$, $t = p, q$; $i = 1, \ldots, n$, $i \neq t$. For $i = t$, $Z_i(t)$ is not defined. The $Z_i(t)$'s are then arranged in an ascending order,

$$Z_{i_1}(t) \leq Z_{i_2}(t) \leq \cdots \leq Z_{i_n}(t), \qquad t = p, q$$

Define

$$S_r(t) = -\sum_{i=1}^{n} |X_i(t)| + 2\sum_{k=1}^{r} |X_{i_k}(t)|, \qquad t = p, q \tag{4}$$

For each t, determine r such that

$$S_{r-1}(t) < 0 \tag{5a}$$

and

$$S_r(t) > 0 \tag{5b}$$

Let u denote the value of r for which inequalities (5a) and (5b) hold for $t = p$ and let v denote the value of r for $t = q$. Then the observation for $t = p$ which corresponds to rank index v is observation p. The estimate b_1 is given by

$$b_1 = Z_{i_u}(p) = Z_q(p) \qquad \text{or}$$

$$= Z_{i_v}(q) = Z_p(q)$$

which is the same as (3b).

From the preceding discussion, it is clear that given the defining observations (X_p, Y_p) and (X_q, Y_q) for the simple MSAE regression, two computations play a major role: the ratios $Z_i(t)$ and the sum $S_r(t)$. For a given t, the rank of $Z_i(t)$ clearly depends on the values of both X_i and

Y_i; however, once the ratios have been computed and arranged in an ascending order, the value of $S_r(t)$ (and thus u and v) depends only on the values of $X_i(t)$'s. Based on these observations, we now develop algorithms to find the intervals on the values of the response variable and the values of the regressor variable within which either the MSAE regression does not change or the defining observations remain the same.

2.1 An Interval for the Value of a Response Variable

The MSAE estimate of β_1 is equal to $Z_{i_r}(t)$, where i_r denotes the rank of the ratio for which inequalities (5a) and (5b) hold. Clearly, change in the value of Y_i does not change the value of $S_r(t)$ as long as the value of i_r for which (5) holds does not change.

For a nondefining observation i, we can determine the interval within which Y_i can lie, ceteris paribus, without changing the MSAE regression by using t equal to either p or q. Thus, without any loss of generality we shall use t = p. Now, if for an observation i, $Z_i(p)$ is less (greater) than b_1, then the estimates b_0 and b_1 do not change as long as the value of Y_i is changed such that the new $Z_i^*(p)$ continues to be less (greater) than b_1. That is, the interval is obtained by solving

$$\frac{Y_i^* - Y_p}{X_i - X_p} \leqslant b_1 \tag{6a}$$

or

$$\frac{Y_i^* - Y_p}{X_i - X_p} \geqslant b_1 \tag{6b}$$

depending on whether $Z_i(p)$ was less than or greater than b_1, where Y_i^* denotes the new value of Y_i.

However, when we consider the defining observation p or q, it is clear that the values of b_0 and b_1 will change for any change in the value of Y_p or Y_q. If we do not want the defining observations to change, we would like to change the value of Y_p (or Y_q) such that the observation p (or q) continues to correspond to u (or v) for which inequalities (5a) and (5b) hold. For observation p this can be accomplished if the new value Y_p^* is such that $Z_p^*(q)$ lies in the interval $[\max(Z_{i_{v-1}}(q), Z_{i_{u-1}}(p)), \min(Z_{i_{v+1}}(q), Z_{i_{u+1}}(p))]$. The value of b_1 will equal $Z_p^*(q)$ and the corresponding value of b_0 can be calculated from equation (3a). Similar statements hold for observation q.

2.2 An Interval for the Value of a Regressor Variable

For an observation i, if we want to determine an interval on the value of the regressor variable, ceteris paribus, within which either the MSAE regression will not change or the defining observations will remain the same, we observe that a change in the value of the regressor variable changes the values of $Z_i(t)$ and $S_r(t)$.

Consider a nondefining observation i. If $Z_i(t) \leqslant b_1$, then X_i^*, the new value of X_i should be such that

$$-\sum_{i=1}^{n} |X_i(t)| - |X_i - X_t| + |X_i^* - X_t| + 2\sum_{k=1}^{r-1} |X_{i_k}(t)| < 0$$

$$-\sum_{i=1}^{n} |X_i(t)| - |X_i - X_t| + |X_i^* - X_t| + 2\sum_{k=1}^{r} |X_{i_k}(t)| \geqslant 0$$

and

$$\frac{Y_i(t)}{X_i^* - X_t} \leqslant b_1$$

On the other hand, if $Z_i(t) \geqslant b_1$, then X_i^* should be such that

$$-\sum_{i=1}^{n} |X_i(t)| + |X_i - X_t| - |X_i^* - X_t| + 2\sum_{k=1}^{r-1} |X_{i_k}(t)| < 0$$

$$-\sum_{i=1}^{n} |X_i(t)| + |X_i - X_t| - |X_i^* - X_t| + 2\sum_{k=1}^{r} |X_{i_k}(t)| \geqslant 0$$

and

$$\frac{Y_i(t)}{X_i^* - X_t} \geqslant b_1$$

The interval which satisfies all these inequalities will be such that if X_i^* lies in it, the MSAE regression will not change.

For a defining observation when the value of the regressor variable is changed, the MSAE estimates b_0 and b_1 change. However, for observation p to continue to be a defining observation, X_p^*, the new value of X_p, should satisfy the following inequalities:

$$-\sum_{i=1}^{n} |X_i(q)| - |X_p - X_q| + |X_p^* - X_q| + 2\sum_{k=1}^{v-1} |X_{i_k}(q)| < 0$$

$$-\sum_{i=1}^{n} |X_i(q)| - |X_p - X_q| + |X_p^* - X_q| + 2\sum_{k=1}^{v} |X_{i_k}(q)| \geq 0$$

and $Y_i(q)/(X_p^* - X_q)$ lies in the interval $[Z_{i_{v-1}}(q), Z_{i_{v+1}}(q)]$. Similar inequalities have to be satisfied for observation q to remain a defining observation.

3. EXAMPLE

The data in Table 1 are taken from Draper and Smith (1981, p. 184). For these data, the MSAE regression line passes through observations numbered 3 and 7, that is, (5.0, 45) and (7.0, 62).

Tables 2 and 3 give the Z_i values and their ranks for t = 3 and t = 7, respectively. For Table 2, $\Sigma_{i=1}^{7} |X_i - X_3| = \Sigma_{i=1}^{7} |X_i(3)| = 6.5$ and for Table 3, $\Sigma_{i=1}^{7} |X_i - X_7| = \Sigma_{i=1}^{7} |X_i(7)| = 10.5$.

TABLE 1 Data for the Example

Observation number, i	X_i	Y_i
1	4.0	33
2	4.5	42
3	5.0	45
4	5.5	51
5	6.0	53
6	6.5	61
7	7.0	62

TABLE 2 The Ratios and Their Ranks for t = 3

Ratio Z_i	Rank	Observation number, i
6.00	1	2
8.00	2	5
8.50	3	7
10.67	4	6
12.00	5	1
12.00	6	4
—	7	3

To construct an interval for the value of a response variable for a nondefining observation, we first consider observation number 2 for which $Z_i(t)$ is less than b_1, in both Tables 2 and 3. Using the information from Table 2,

$$\frac{Y_2^* - 45}{4.5 - 5.0} \leqslant 8.50$$

TABLE 3 The Ratios and Their Ranks for t = 7

Ratio Z_i	Rank	Observation number, i
2.00	1	6
7.33	2	4
8.00	3	2
8.50	4	3
9.00	5	5
9.67	6	1
—	7	7

or

$$Y_2^* \geqslant 40.75$$

That is, as long as the value of the response variable for observation number 2 is greater than or equal to 40.75, the MSAE regression line will not change.

Now we consider observation number 5 for which $Z_5(3) < b_1$ and $Z_5(7) > b_1$. Using the information from Table 2,

$$\frac{Y_5^* - 45}{6.0 - 5.0} \leqslant 8.50$$

or

$$Y_5^* \leqslant 53.50$$

That is, the MSAE regression line will not change if the value of the response variable for observation 5 is less than or equal to 53.50.

Now we consider the defining observations. Consider observation number 7. To obtain the interval within which the value of the response variable for observation number 7 can lie such that observation number 7 will continue to be a defining observation, Y_7^* has to satisfy the following inequalities:

$$8.0 \leqslant \frac{Y_7^* - 45}{7.0 - 5.0} \leqslant 9.0$$

or

$$61.00 \leqslant Y_7^* \leqslant 63.00$$

In other words, observation number 7 will continue to be a defining observation if Y_7^* lies in the interval [61.00, 63.00].

The intervals on the values of the response variable for the example are summarized in Table 4.

To construct an interval for the value of a regressor variable for a nondefining observation, we first consider observation number 2 for which $Z_2(t)$ is less than b_1 in both Tables 2 and 3. For Table 2, X_2^* should satisfy the following inequalities:

TABLE 4 Intervals for the Values of the Response Variable and the Regressor Variable

Observation number, i	Interval for Y_i	Interval for X_i
1	$(-\infty, 36.50]$	$[3.588, 5.500]$
2	$[40.75, \infty)$	$[3.000, 4.647]$ and $[7.000, 9.000]$
3	$[44.00, 46.00]$	$[4.875, 5.111]$
4	$[49.25, \infty)$	$[4.000, 5.706]$ and $[6.000, 8.000]$
5	$(-\infty, 53.50]$	$[5.941, 7.500]$
6	$[57.75, \infty)$	$[5.000, 7.000]$
7	$[61.00, 63.00]$	$[6.593, 7.125]$

$$-6.5 - 0.5 + |X_2^* - 5.0| + 2(3.5) \geq 0$$

$$-6.5 - 0.5 + |X_2^* - 5.0| + 2(1.5) < 0$$

and

$$\frac{-3.00}{X_2^* - 5.0} \leq 8.50$$

For Table 3, X_2 should satisfy the following inequalities:

$$-10.5 - 2.5 + |X_2^* - 7.0| + 2(6.5) \geq 0$$

$$-10.5 - 2.5 + |X_2^* - 7.0| + 2(4.5) < 0$$

and

$$\frac{-20.00}{X_2^* - 7.0} \leq 8.50$$

Solving these inequalities, we get two disjoint intervals (3.000, 4.647] and (7.000, 9.000) for the value of X_2^*. That is, if the value of the regressor variable lies in either the interval (3.000, 4.647] or the interval (7.000, 9.000), the MSAE regression line for the example will not change.

Next we consider observation number 5 for which $Z_5(3) < b_1$ and $Z_5(7) > b_1$. Using the information from Table 2, X_5^* should satisfy the following inequalities:

$$-6.5 - 1.0 + |X_5^* - 5.0| + 2(3.5) \geq 0$$

$$-6.5 - 1.0 + |X_5^* - 5.0| + 2(1.5) < 0$$

and

$$\frac{8.00}{X_5^* - 5} \leq 8.50$$

For Table 3, X_5^* should satisfy the following inequalities:

$$-10.5 + 1.0 - |X_5^* - 7.0| + 2(6.5) \geq 0$$

$$-10.5 + 1.0 - |X_5^* - 7.0| + 2(4.5) < 0$$

and

$$\frac{-9.00}{X_5^* - 7} \geq 8.50$$

The solution to these simultaneous equations given an interval [5.941, 7.000) for the value of X_5^*. That is, if the value of the regressor variable for observation 5 lies in the interval [5.941, 7.000), the MSAE regression equation for the data will not change.

Now we consider the defining observations. For observation number 7, X_7^* should satisfy the following inequalities:

$$-6.5 - 2.0 + |X_7^* - 5.0| + 2(3.5) \geq 0$$

$$-6.5 - 2.0 + |X_7^* - 5.0| + 2(1.5) < 0$$

and

$$8.00 \leq \frac{17}{X_7^* - 5.0} \leq 10.67$$

These inequalities are satisfied simultaneously when X_7^* lies in the interval [6.500, 7.125]. If the value of the regression variable for observation 7 lies in the interval [6.500, 7.125], then observation number 7 will continue to be a defining observation.

The intervals for the values of the regressor variable for the example are summarized in Table 4.

4. CONCLUDING REMARKS

In regression analysis and, in fact, in most statistical analyses, it is tacitly assumed that all the observations have been taken accurately and precisely. If a certain observation does not confirm to the model, it is considered an outlier. It is concluded that it is due to a recording or a transmission error in the observation or that the observation represents an unusual data point and as such needs further investigation. These may be valid conclusions. The quality of the remaining observations is never questioned or checked. It is possible that the remaining observations are off because of the bias in the instruments or because the observations were taken on different days, at different places, or by different individuals.

In this chapter, we have shown how to determine intervals on the value of a response variable and the value of a regressor variable within which their values can vary, ceteris paribus, such that either the MSAE regression does not change or the defining observations for it remain the same. This analysis provides very useful information to the investigator. For example, if for certain observations the interval is "narrow," it will be useful to ensure that the observation had been properly taken, recorded, and transmitted. This analysis allows the user to examine critically a data set and the MSAE regression fit and thus have more confidence in the results and conclusions.

REFERENCES

Barrodale, I. (1968). L_1 approximation and the analysis of data. Appl. Stat. 17: 51-56.

Beckman, R. J., and Cook, R. D. (1983). Outliers. Technometrics 25: 119-149.

Brownlee, K. A. (1965). Statistical Theory and Methodology in Science and Engineering. Wiley, New York.

Dielman, T., and Pfaffenberger, R. (1982). LAV (least absolute value) estimation in linear regression: A review. TIMS Stud. Manage. Sci. 19: 31-52.

Dielman, T., and Pfaffenberger, R. (1984). Computational algorithms for calculating least absolute value and Chebyshev estimates for multiple regression. Am. J. Math. Manage. Sci. 4: 169-197.

Draper, N. R., and Smith, H. (1981). Applied Linear Regression, 2d ed. Wiley, New York.

Hampel, F. R. (1971). A general qualitative definition of robustness. Ann. Math. Stat. 42: 1887-1896.

Huber, P. J. (1974). Comment on "Adaptive robust regression: A partial review and some suggestions for future applications and theory," by R. V. Hogg. J. Am. Stat. Assoc. 69: 926-927.

Karst, O. J. (1958). Linear curve fitting using least deviations. J. Am. Stat. Assoc. 53: 118-132.

Narula, S. C. (1982). Optimization techniques in regression analysis: A review. TIMS Stud. Manage. Sci. 19: 11-29.

Narula, S. C. (1987). The minimum sum of absolute errors regression. J. Qual. Technol. 19: 37-45.

Narula, S. C., and Wellington, J. F. (1982). The minimum sum of absolute errors regression: A state of the art survey. Int. Stat. Rev. 50: 317-326.

Narula, S. C., and Wellington, J. F. (1985). Interior analysis for the minimum sum of absolute errors regression. Technometrics 27: 181-188.

Rice, J. R., and White, J. S. (1964). Norms for smoothing and estimation. SIAM Rev. 6: 243-256.

Sposito, V. A., Smith, W. C., and McCormick, C. (1978). Minimizing the Sum of Absolute Deviations. Vandenhoeck & Ruprecht, Gottingen, West Germany.

7

Residuals in Variance-Component Models

WILLIAM H. FELLNER E. I. du Pont de Nemours and Company, Inc., Wilmington, Delaware

1. INTRODUCTION

In regression analysis, the residual is the basis for a wide variety of robust and diagnostic methods. Observations with large residuals can be iteratively winsorized or downweighted in order to produce an outlier-resistant analysis. Moreover, patterns in the residuals can reveal unexpected trends, lack of normality, and/or heterogeneity of variance.

The residual estimates the random error in the corresponding observation. In fact, it is the best linear unbiased estimate of that error.

In the variance-component model, each observation is formed from the sum of several random errors. A basis for extending robust and diagnostic methods to this model is to obtain residuals for each of the random errors.

This chapter proposes the use of best linear unbiased estimates in order to estimate both the fixed parameters and the random errors in the mixed linear model (Henderson, 1963). These estimates depend on the ratios of the variance components. Given these components, the estimates can be expressed as the solution to a regression problem. Thus, well-known robust and diagnostic methods extend in a straightforward manner.

The best linear unbiased estimates are intimately tied to estimation of the variance components by ordinary or restricted maximum likelihood. Harville (1977) gives an extensive review of these methods. If the variance components are unknown, then iterative solution of the above regression problem yields simultaneous estimates of the fixed parameters, the residuals, and the variance components.

This chapter reviews Henderson's estimates, estimates of the variance components based on likelihood methods, and a robust extension of these methods (Fellner, 1986). Analyses of two data sets are used to illustrate the concepts.

2. THE BALANCED ONE-WAY MODEL

Consider first the balanced, random one-way model

$$y_{ij} = \mu + a_i + e_{ij} \qquad (i = 1,\ldots,I;\ j = 1,\ldots,J)$$

where $a_i \sim N(0,\sigma_1^2)$, $e_{ij} \sim N(0,\sigma_2^2)$, and the a's and e's are mutually independent. Let α_i and ϵ_{ij} be the "realized values" of a_i and e_{ij}, respectively.

Let estimates of μ, α_i, and ϵ_{ij} be denoted by $\hat{\mu}$, $\hat{\alpha}_i$, and $\hat{\epsilon}_{ij}$, respectively. Then α_i and $\hat{\epsilon}_{ij}$ are the residuals for the data y_{ij}.

Traditionally, residuals for the balanced one-way design are obtained using

$$\hat{\mu} = y_{..}$$

$$\hat{\alpha}_i = y_{i.} - y_{..}$$

and

$$\hat{\epsilon}_{ij} = y_{ij} - y_{i.}$$

where $y_{i.}$ is the average of the observations in the ith group, and $y_{..}$ is the average of the $y_{i.}$'s.

These estimates can be described as the solution to a least squares problem as follows. Subject to the constraint

$$y_{ij} = \hat{\mu} + \hat{\alpha}_i + \hat{\epsilon}_{ij}$$

minimize $\Sigma\Sigma\hat{\epsilon}_{ij}^2$. Then, given the resulting values for $\hat{\epsilon}_{ij}$, minimize $\Sigma\alpha_i^2$.

Viewed this way, the method extends naturally to unbalanced nested designs, as well as unbalanced crossed designs with no missing cells. It is intimately related to the "unweighted means analysis" of Yates (1934), as described by Searle (1971, Section 8.3c). It will be convenient to call these estimates the unweighted means estimates.

These estimates may exhibit some inconsistencies. When σ_1^2 is small in comparison with σ_2^2, the values of $\hat{\alpha}$ will appear large relative to σ_1. Moreover, the unweighted means estimates become ambiguous when applied to crossed designs with missing cells.

If σ_1^2 and σ_2^2 are known, Henderson's (1963) approach is to minimize

$$\sigma_1^{-2}\sum\hat{\alpha}_i^2+\sigma_2^{-2}\sum\sum\hat{\epsilon}_{ij}^2$$

This yields the estimates

$$\hat{\mu}=y_{..}$$

$$\hat{\alpha}_i=\frac{(y_{i.}-y_{..})\sigma_1^2}{\sigma_1^2+\sigma_2^2/J}$$

and

$$\hat{\epsilon}_{ij}=y_{ij}-y_{..}-\hat{\alpha}_i$$

It is seen that the unweighted means residuals are obtained by assuming $\sigma_1^2 \gg \sigma_2^2$.

Henderson's estimates have a number of advantages over the unweighted means estimates. They are best linear unbiased estimates. They are consistent with the values of the variance components: when $\sigma_1^2 = 0$, the $\hat{\alpha}_i$ are all zero. Finally, they can be extended to crossed models with arbitrary numbers of missing cells.

On the other hand, because these estimates depend on assumed values of the variance components, they are most suitable for large data sets from which reasonably precise estimates of the variance components are available.

Since the estimates are linear, standard errors are readily obtained:

$$\text{S.E.}^2(\hat{\mu}) = I^{-1}\sigma_1^2 + (IJ)^{-1}\sigma_2^2$$

$$\text{S.E.}^2(\hat{\alpha}_i) = \left\{\frac{(I-1)J\sigma_1^2}{IJ\sigma_1^2 + I\sigma_2^2}\right\}\sigma_1^2 \tag{1}$$

$$\text{S.E.}^2(\hat{\epsilon}_{ij}) = \left\{\frac{IJ(J-1)\sigma_1^2 + (IJ-1)\sigma_2^2}{IJ^2\sigma_1^2 + IJ\sigma_2^2}\right\}\sigma_2^2 \tag{2}$$

From a diagnostic point of view, the standard errors serve as a basis for standardizing the residuals, $\hat{\alpha}_i$ and $\hat{\epsilon}_{ij}$

As in regression, the squared standard error of each residual is equal to the corresponding variance multiplied by a factor between zero and one. In regression, one minus this factor is commonly called the <u>leverage</u>. A residual with a high leverage represents an observation with an outlying combination of regressor values or, in the case of weighted regression, a heavy weight. Such an observation has a strong influence on the fit of the model at its combination of regressor values. As a result, the magnitude of the residual greatly underestimates the true error in the observation.

In the present context, the leverages are

$$\text{lev}(\hat{\alpha}) = \frac{J\sigma_1^2 + I\sigma_2^2}{IJ\sigma_1^2 + I\sigma_2^2}$$

for each element of $\hat{\alpha}$, and

$$\text{lev}(\hat{\epsilon}) = \frac{IJ\sigma_1^2 + \sigma_2^2}{IJ^2\sigma_1^2 + IJ\sigma_2^2}$$

for each element of $\hat{\epsilon}$. It is seen that residuals associated with small variance components are given heavy weights and thus large leverages.

In practice, σ_1^2 and σ_2^2 are not known and must be estimated from the data. The method of restricted maximum likelihood (Patterson and Thompson, 1971) is intimately associated with Henderson's estimates. In fact, the restricted maximum likelihood estimates can be written

$$\hat{\sigma}_1^2 = \frac{\sum \hat{\alpha}_i^2}{I\{1 - \mathrm{lev}(\hat{\alpha})\}}$$

and

$$\hat{\sigma}_2^2 = \frac{\sum\sum \hat{\epsilon}_{ij}^2}{IJ\{1 - \mathrm{lev}(\hat{\epsilon})\}}$$

where the $\hat{\alpha}$, $\hat{\epsilon}$, and their leverages are computed from $\hat{\sigma}_1^2$ and $\hat{\sigma}_2^2$. In general, restricted maximum likelihood estimates of the variance components must be obtained iteratively. In the present context, and for balanced data generally, they are equivalent to the usual estimates derived from mean squares:

$$\hat{\sigma}_1^2 = J^{-1}(\mathrm{MSA} - \mathrm{MSE})$$

$$\hat{\sigma}_2^2 = \mathrm{MSE}$$

for MSA $\geqslant$ MSE, and

$$\hat{\sigma}_1^2 = 0$$

$$\hat{\sigma}_2^2 = \mathrm{MST}$$

for MSA < MSE. Here MSA, MSE, and MST are the mean squares among groups, within groups and total, respectively (Thompson, 1962) as described by Searle (1971, Section 9.9g).

When estimates of the variance components are used to obtain $\hat{\alpha}$ and $\hat{\epsilon}$, equations (1) and (2) understate the true standard errors of these quantities, since $\hat{\alpha}$ and $\hat{\epsilon}$ are dependent on the variance-component estimates. Similarly, the denominators of $\hat{\sigma}_1^2$ and $\hat{\sigma}_2^2$ are not degrees of freedom in the classical sense. Rather, if the numerators are regarded as approximations of scaled chi-square variates, the denominators give "degrees of freedom" that are too large.

3. EXAMPLE: PAINT CAN EAR THICKNESS DATA

Snee (1983) analyzed the thicknesses of 150 paint can ears, sampled over a period of 30 days. To emphasize the diagnostic value of Henderson's estimates, let us consider only the first three samples of each day. The data are shown in Table 1, together with the resulting variance-component estimates. The day-to-day variance represents 13% of the

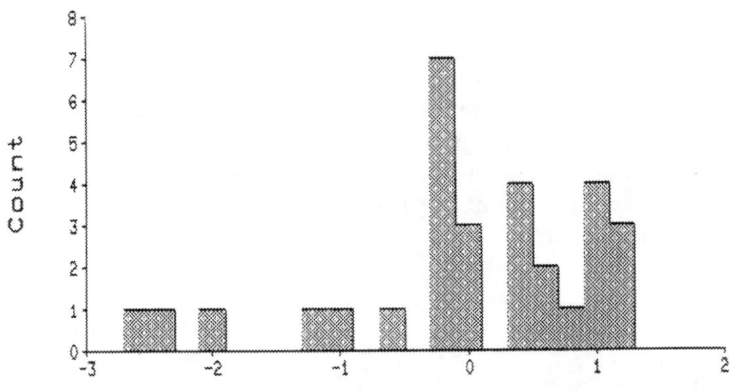

FIGURE 1 Paint can ear thickness data: histogram of standardized DAY residuals, obtained by Henderson's method.

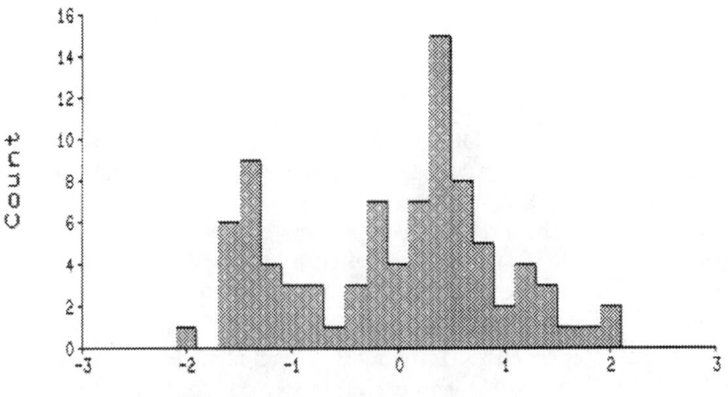

FIGURE 2 Paint can ear thickness data: histogram of standardized EAR residuals, obtained by Henderson's method.

TABLE 1 Paint Can Ear Thickness Data[a]

Day	Measured thicknesses			Day	Measured thicknesses		
1	29	36	39	16	35	30	35
2	29	29	28	17	40	31	38
3	34	34	39	18	35	36	30
4	35	37	33	19	35	34	35
5	30	29	31	20	35	35	31
6	34	31	37	21	32	36	36
7	30	35	33	22	36	37	32
8	28	28	31	23	29	34	33
9	32	36	38	24	36	36	35
10	35	30	37	25	36	30	35
11	35	30	35	26	35	30	29
12	38	34	35	27	35	36	30
13	34	35	33	28	35	30	36
14	40	35	34	29	38	36	35
15	34	35	38	30	30	34	40

Variance Components

Source	Variance	Percent of total
DAY	1.3	13
OBSERVATION	8.2	87

[a] In thousandths of an inch. Only the first three observations of each day are included.
Source: Snee (1983).

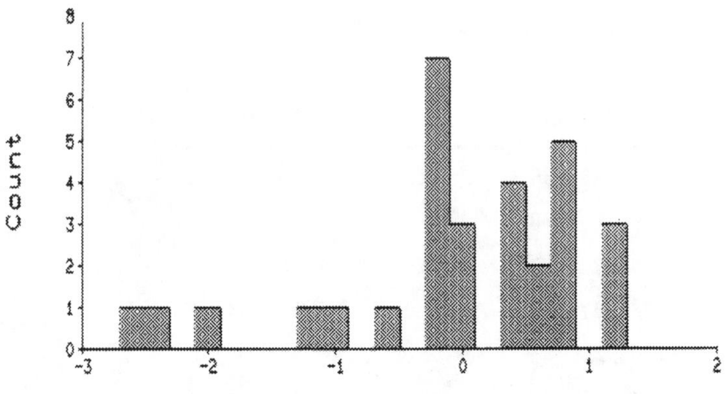

FIGURE 3 Paint can ear thickness data: histogram of standardized DAY residuals, obtained by the method of unweighted means.

total variance. As a result, the leverage for the DAY residuals is 0.69. The leverage for the OBSERVATION residuals is 0.11.

Figure 1 shows a histogram of the DAY residuals, obtained by Henderson's method, standardized with respect to their "nominal" standard errors as given by equation (1) (substituting the estimates of the variance components). Figure 2 shows a histogram of the residuals for individual EARs, similarly obtained and standardized using equation (2).

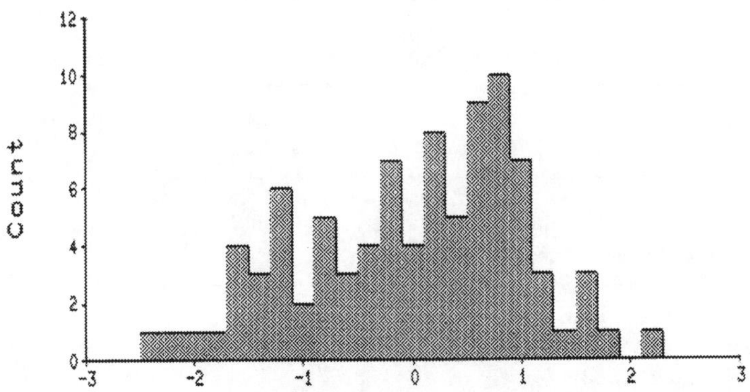

FIGURE 4 Paint can ear thickness data: histogram of standardized EAR residuals, obtained by the method of unweighted means.

What is striking about Figure 2 is the obvious bimodality of these residuals, a phenomenon that Snee noted for the observations taken as a whole. This bimodality was ultimately traced to two machines which were producing paint can ears of different average thicknesses deposited in a common hopper.

Figure 3 shows a histogram of DAY residuals, also standardized, obtained by the method of unweighted means. Figure 4 shows the standardized EAR residuals obtained by this method. The bimodality in Figure 4 is not nearly as striking. The reason is that the EAR residuals are constrained by the requirement that the average of each group of three be zero. In terms of leverage, the "classical" method gives each of the OBSERVATION residuals an effective leverage value of 0.33, considerably higher than the 0.11 accorded by the Henderson method.

Thus, choosing a set of residuals that are consistent with the estimated variance components can greatly improve the diagnostic information contained in the residuals.

4. THE GENERAL MIXED MODEL

The concepts developed for the balanced one-way model extend in a straightforward manner to more complex models. Consider the general linear model

$$y = X\alpha + Z_1 b_1 + \cdots + Z_c b_c + e$$

where y is an $n \times 1$ vector of observations, X is a full-rank $n \times p$ known matrix, α is an unknown order-p vector of fixed effects, Z_i ($i = 1, \ldots, c$) is an $n \times q_i$ known matrix, b_i is an order-q_i vector of random effects, and e is an order-n vector of random errors.

Assume that the b_i and e are independently normally distributed, with $E(b_i) = 0$, $\mathrm{Var}(b_i) = \sigma_i^2 I_i$ ($i = 1, \ldots, c$), and $E(e) = 0$, $\mathrm{Var}(e) = \sigma_{c+1}^2 I_{c+1}$. Here I_i is the order-q_i identity matrix and I_{c+1} is the order-n identity matrix.

Let $\beta_1, \ldots, \beta_c$ and ϵ be the realized values of $b_1, \ldots, b_c$ and e, respectively. Then Henderson's estimates of α, $\beta_1, \ldots, \beta_c$ and ϵ are obtained by minimizing

$$\left(\sum_{i=1}^{c} \sigma_i^{-2} \|\hat{\beta}_i\|^2 \right) + \sigma_{c+1}^{-2} \|\hat{\epsilon}\|^2$$

subject to the constraint

$$y = X\hat{\alpha} + Z_1\hat{\beta}_1 + \cdots + Z_c\hat{\beta}_c + \hat{\epsilon}$$

Henderson's estimates can be written in more compact form. Let Z be the partitioned matrix

$$Z = [Z_1, \ldots, Z_c]$$

and let $\hat{\beta}$ be the partitioned vector

$$\hat{\beta}' = [\hat{\beta}_1', \ldots, \hat{\beta}_c']$$

Let D and R be defined respectively by

$$D = \mathrm{diag}[\sigma_1^2 I_1, \ldots, \sigma_c^2 I_c]$$

and

$$R = \sigma_{c+1}^2 I_{c+1}$$

Then Henderson's estimates for α and $\beta_1, \ldots, \beta_c$ are the least squares solution to

$$C\begin{bmatrix}\hat{\alpha}\\ \hat{\beta}\end{bmatrix} = \begin{bmatrix}R^{-1/2}X & R^{-1/2}Z\\ 0 & D^{-1/2}\end{bmatrix}\begin{bmatrix}\hat{\alpha}\\ \hat{\beta}\end{bmatrix} = \begin{bmatrix}R^{-1/2}y\\ 0\end{bmatrix} \quad (3)$$

with $\hat{\epsilon}$ obtained by

$$\hat{\epsilon} = y - X\hat{\alpha} - Z\hat{\beta} \quad (4)$$

The leverage of an element of $\hat{\beta}$ or $\hat{\epsilon}$ is the corresponding diagonal element of $C(C'C)^{-1}C'$, and the standard error of this element is $[\sigma^2(1 - \text{lev})]^{1/2}$ where σ^2 is the variance-component value appropriate to that element.

As in the case of the balanced one-way model, the restricted maximum likelihood estimates of the variance components are intimately tied to Henderson's estimates. These estimates can be described as follows.

Let $\hat{\sigma}_1^2, \ldots, \hat{\sigma}_{c+1}^2$ represent these estimates, and assume that Henderson's estimates of β and ϵ, together with their leverages, are formed from these estimates. Then, for $i = 1, \ldots, c$, the restricted maximum likelihood estimates satisfy

$$\hat{\sigma}_i^2 = \frac{||\hat{\beta}_i||^2}{\sum \{1 - \text{lev}(\cdot)\}} \tag{5}$$

where the summation is over the elements of $\hat{\beta}_i$; also,

$$\hat{\sigma}_{c+1}^2 = \frac{||\hat{\epsilon}||^2}{\sum \{1 - \text{lev}(\cdot)\}} \tag{6}$$

where the summation is over the elements of $\hat{\epsilon}$.

Note that the denominators in (5) and (6) sum to $n - p$. Thus, it is not necessary to compute the individual leverages of the elements of $\hat{\epsilon}$ on each iteration.

The use of equations (5) and (6) as a basis for iteratively computing the variance component estimates was suggested by Harville (1977, Section 6).

The calculations described above have the advantage of not requiring the off-diagonal elements of $(C'C)^{-1}$ or of $C(C'C)^{-1}C'$. As noted by Fellner (1987), in the vast majority of applications the matrix $C'C$ is sparse; that is, it has very few nonzero elements. After suitable rearrangement of the columns of C, its Cholesky factor will be sparse as well. Only the nonzeros need to be explicitly stored or used in computation. By contrast, $(C'C)^{-1}$ and $C(C'C)^{-1}C'$ will generally be full matrices, and their complete calculation will require orders of magnitude more computational effort. For this reason as well, only diagnostics that do not require the off-diagonal elements of these matrices are considered in this chapter.

5. THE GENERAL MIXED MODEL: ROBUST ESTIMATION

Outlier-resistant methods can be applied to any method for calculating residuals. The basic idea is to downweight or winsorize the residuals that are unusually large. In this manner, Rocke (1983) and Shoemaker (1980) developed robust variance-component estimates roughly based on the unweighted means residuals. By contrast, the method of Fellner (1986) uses the Henderson residuals. We will review the latter as an example of the construction of outlier-resistant methods from corresponding "classical" methods.

Fellner's method is a straightforward extension of Huber's (1964) outlier-resistant approach to regression. Let the function $\psi(u)$ be odd, monotonic, and bounded by $\pm r$. As an example, Huber suggested the function

$$\psi(u) = \min\{r, \max(-r, u)\}$$

In place of Henderson's estimates, Fellner obtains estimates using

$$C\begin{bmatrix} \hat{\alpha} \\ \hat{\beta} \end{bmatrix} = \begin{bmatrix} R^{-1/2}X & R^{-1/2}Z \\ 0 & D^{-1/2} \end{bmatrix}\begin{bmatrix} \hat{\alpha} \\ \hat{\beta} \end{bmatrix} = \begin{bmatrix} R^{-1/2} & \tilde{y} \\ D^{-1/2} & \tilde{0} \end{bmatrix} \tag{7}$$

with $\hat{\epsilon}$ still determined by equation (4). Here

$$\tilde{y} = X\hat{\alpha} + Z\hat{\beta} + R^{1/2}\psi(R^{-1/2}\hat{\epsilon}) \tag{8}$$

and

$$\tilde{0} = \hat{\beta} - D^{1/2}\psi(D^{-1/2}\hat{\beta}) \tag{9}$$

where the function ψ is applied element by element. These are "pseudo-observations" in the sense of Bickel (1976, p. 176). If the function ψ has been selected appropriately, then except for outlying elements of $\hat{\epsilon}$ or $\hat{\beta}$, the elements of $\tilde{y}$ and $\tilde{0}$ will be close to y and zero, respectively.

When elements of $\hat{\epsilon}$ and $\hat{\beta}$ are outlying, the corresponding $\tilde{y}$ and $\tilde{0}$ are closer to $X\hat{\alpha} + Z\hat{\beta}$ and $-\hat{\beta}$, respectively. In effect, the corresponding observed values on the right side of equation (7) are moved closer to their predicted values, reducing the tendency for these elements of $\hat{\beta}$ and $\hat{\epsilon}$ to be kept small.

To obtain outlier-resistant estimates of the variance components, equations (5) and (6) are modified as follows:

$$\hat{\sigma}_i^2 = \frac{\hat{\sigma}_i^2 ||\psi(\hat{\beta}_i/\hat{\sigma}_i)||^2}{h\sum\{\hat{1} - \text{lev}(\cdot)\}} \tag{10}$$

and

$$\hat{\sigma}_{c+1}^2 = \frac{\hat{\sigma}_{c+1}^2 ||\psi(\hat{\epsilon}/\hat{\sigma}_{c+1})||^2}{h\sum\{1 - \text{lev}(\cdot)\}} \tag{11}$$

where, again, the function ψ is applied element by element, and where the denominators are defined as in equations (5) and (6), except for the constant h. This constant given by

$$h = E_\Phi \psi^2(X)$$

where Φ denotes that the expectation is taken over a standard normal distribution. Under normality, the constant h reduces the bias in the estimates of the large variance components which was introduced by applying the function ψ.

Fellner evaluates equations (4) and (7) through (11) iteratively for the estimated quantities.

As noted by Andrews (1979) for regression models, outlier-resistant methods, precisely because of their insensitivity to outliers, make any outliers in the data more visible when the residuals are used for diagnostics. Other departures from assumptions become more visible as well, because of the limited influence of outlying residuals on the other residuals (Fellner, 1986). This becomes apparent in the example which follows.

6. EXAMPLE: BULK PRODUCT PHYSICAL PROPERTY

The data come from a special study of the sources of variation of a physical property measured on a bulk product. From each production lot, two samples were obtained from the first "package." Each sample was tested by each of two different analysts. From the same production lot, a single sample was obtained from a later "package." This sample was tested by the first of the two analysts. In this way, 93 production lots were sampled and tested. Nominally, a different pair of analysts were used for each of the production lots. The data are shown in Table 2. Missing values are left blank.

TABLE 2 Bulk Product Physical Property: Data

	Package: 1	1	1	1	2
	Sample: 1	1	2	2	1
	Analyst: 1	2	1	2	1
Lot					
1	21.83	21.94	21.04	18.45	20.15
2	24.38	24.66	19.90	20.58	23.43
3	22.35	22.94	22.34	23.82	23.90
4	27.62	26.07	24.64	24.63	24.38
5	22.49	23.95	23.48	22.86	22.41
6	15.55	15.75	23.92	24.99	25.70
7	25.11	25.19	26.60	27.20	23.14
8	23.00	22.91	27.44	28.23	22.39
9	23.44		24.83		23.05
10	25.71	25.25	25.73	26.37	26.68
11	24.83	25.81	17.42	17.54	15.09
12	24.38	23.81	19.67	19.02	21.38
13	22.74	24.95	20.55	17.81	25.16
14	22.85	22.26	24.17	23.19	22.95
15	25.53	23.93	19.14	15.65	23.47
16	23.24	24.18	24.70	24.77	23.88
17	23.67	24.39	22.01	21.10	22.93
18	24.02	24.30	21.08	20.29	21.04
19	23.82	23.89	21.01	21.72	22.18
20	23.81	24.74	23.51	23.14	21.28
21	25.32	25.22	24.15	23.61	24.53
22	24.81	25.33	23.42	24.57	11.75*
23	23.90	24.95	24.29	26.36	19.54
24	26.51	27.52	25.13	24.71	23.98
25	26.39		23.74		22.76
26	27.78	25.21	12.21*	18.12	15.01
27	26.30	27.06	28.36	26.13	26.47
28	24.56	24.49	26.35	25.55	25.44
29	23.37	22.79	25.56	26.20	25.84
30	26.04	26.62	21.26	15.20	18.12
31	21.67	25.60	23.13	26.04	25.70

TABLE 2 (continued)

Package:	1	1	1	1	2
Sample:	1	1	2	2	1
Analyst:	1	2	1	2	1
Lot					
32	26.66	25.52	26.12	25.57	24.88
33	26.71	26.60	23.38	23.28	20.58
34	25.72	25.45	25.28	25.08	25.72
35	22.16	25.55	24.65	24.33	22.83
36	27.30	28.64	25.70	25.87	24.35
37	26.76	25.97	25.61	25.26	24.81
38	25.71	25.99	23.87	23.02	24.03
39	26.44	25.59	24.41	23.25	24.05
40	24.35	26.66	22.85	25.96	26.10
41	26.86	27.10	24.90	25.06	26.86
42	24.95	24.84	25.20	27.42	23.63
43	23.82	22.73	23.10	22.79	22.27
44	23.81	22.75	22.15	21.92	
45	24.63	26.62	25.54	27.12	24.04
46	26.10	26.10	25.75	26.19	26.50
47	25.02	26.72	27.42	26.71	27.10
48	26.22	28.23	24.48	26.52	25.87
49	26.91	26.61	26.16	26.93	23.62
50	29.09	29.28	28.34	28.55	28.00
51	27.16	25.99	26.36	27.18	17.41
52	26.69	26.66	24.71	23.29	27.05
53	22.69	22.93	22.42	22.41	22.44
54	26.03	25.42	21.44	24.53	15.41
55	24.06	24.09	23.09	23.03	24.30
56	24.85	25.57	26.11	26.60	26.76
57	26.41	26.87	24.97	24.56	24.87
58	24.50	25.01	20.80	21.37	24.51
59	27.19	27.31	24.40	26.73	25.34
60	26.56	26.84	25.85	26.45	26.70
61	24.77	23.03	10.13*	10.76*	13.13*
62	25.12	21.01	11.06*	11.17*	21.82

(continued)

TABLE 2 (continued)

Package: 1 Sample: 1 Analyst: 1	1 1 2	1 2 1	1 2 2	2 1 1
Lot				
63	24.61	26.40	21.03	22.19	22.51
64	23.97	24.55	13.76*	13.49*	22.94
65	24.90	24.93	23.28	22.65	21.68
66	24.33	25.44	21.67	22.45	21.50
67	21.40	21.22	21.70	22.66	22.16
68	21.30	22.31	19.66	19.12	20.36
69	24.88	27.02	12.61*	10.83*	24.13
70	24.32	25.85	24.32	24.16	24.01
71	25.71	25.25	27.11	27.13	24.52
72	25.03	24.50	24.56	26.01	24.41
73	26.44	26.80	22.75	24.65	23.65
74	25.05		25.15		17.81
75	24.12	24.66	24.46	25.06	24.36
76	24.91	25.00	21.37	21.81	23.20
77	24.04	22.47	19.73	19.49	21.18
78	19.08	19.66	25.09	24.25	22.55
79	23.26	23.31	22.80	23.71	22.85
80	13.65*	11.22*	25.66	18.07	24.65
81	23.70	26.02	24.40	24.39	24.57
82	26.27	26.28	25.31	24.64	24.18
83	25.68	25.75	23.08	25.80	24.41
84	20.93	16.22	27.02	28.25	25.45
85	22.93	24.16	21.10	22.36	23.20
86	26.54	26.85	24.45	24.61	26.45
87	26.37		26.00		23.84
88	26.64	26.31	26.26	26.00	25.64
89	26.15	26.53	23.71	25.18	24.61
90	22.05	24.67	26.84	23.56	28.24
91	24.79	28.13	26.10	25.61	24.66
92	27.75	26.80	16.79	26.60	26.13
93	26.11	26.24	25.02	24.96	23.50

A scan of Table 2 shows a number of low outliers. Those less than 14 are marked with an asterisk. Note that the low outliers tend either to come in pairs from the same sample or to come from package 2. This suggests that the low outliers are mostly outlying samples. Also note that lot 61 had two such samples. For this lot, it is difficult to know whether the lot average is low or the two low samples are outliers.

The following mixed model was fit to the data:

$$y_{ijkl} = \mu_{jkl} + a_i + b_{ij} + c_{ijk} + d_{il} + e_{ijkl}$$

where

μ_{jkl} is the fixed mean of the (jkl)th combination of package, sample, and analyst.

a_i is the random effect of the ith lot (LOT effect).

b_{ij} is the random effect of the jth package from the ith lot (PACKAGE effect).

c_{ijk} is the random effect of the kth sample from the jth package from the ith lot (SAMPLE effect).

d_{il} is the random effect of the lth analyst for the ith lot (ANALYST effect).

e_{ijkl} is the random effect of the (ijkl)th observation (MEASUREMENT effect).

Both Fellner's outlier-resistant method (robust) and the restricted maximum likelihood method (nonrobust) were used to analyze the data. The robust method utilized Huber's ψ function with $r = 2$.

Note that the method of unweighted means cannot be used to analyze these data, since package and analyst are crossed and there are missing cells.

Table 3 shows the means for the five combinations of package, sample, and analyst, averaged over all 93 lots. The means from the robust analysis are slightly higher because of its resistance to the low outliers. The two means for package 1, sample 1 are higher than the other three. This bias was ultimately traced to the fact that the first sample of the first package usually came from the start of the production lot, before process equilibrium was achieved.

Table 4 shows the estimates of the variance components. The variances for PACKAGE and ANALYST are virtually zero. SAMPLE variance is more than half the total variance. LOT variance accounts for another third or so. The remaining sixth is what we have called MEASUREMENT variance.

TABLE 3 Bulk Product Physical Property: Means

Package	Sample	Analyst	Mean (robust)	Mean (nonrobust)
1	1	1	24.8	24.6
1	1	2	25.0	24.8
1	2	1	23.6	23.2
1	2	2	23.7	23.4
2	1	1	23.4	23.3

The nonrobust SAMPLE and MEASUREMENT variances are considerably larger than the corresponding robust variances. Again, this is due to the outlier resistance of the robust method.

Except for the small number of missing values, the same sampling plan was followed for each lot. As a result, the leverages for corresponding residuals of different lots are largely equal. Table 5 shows this pattern of leverages. The leverages depend on the variance components. Thus the PACKAGE and ANALYST leverages are close to one. The leverage for the SAMPLE residual from package 2 is about twice that of a SAMPLE residual from package 1. This is because only one measurement is made on the package 2 sample.

The single observation made on the sample from package 2 has two residuals associated with it: a SAMPLE residual and a MEASUREMENT

TABLE 4 Bulk Product Physical Property: Variance Components

Source	Variance (robust)	Variance (nonrobust)
Lot	2.2	2.3
Package	0.1	0.1
Sample	3.4	5.8
Analyst	0.0	0.1
Measurement	1.0	1.4

TABLE 5 Bulk Product Physical Property: Leverages[a]

Lot residuals

0.51

Package residuals

Package 1: 0.99
Package 2: 0.99

Sample residuals

Package 1, sample 1: 0.28
Package 1, sample 2: 0.28
Package 2, sample 1: 0.34

Analyst residuals

Analyst 1: 0.95
Analyst 2: 0.95

Measurement residuals

Package 1, sample 1, analyst 1: 0.48
Package 1, sample 1, analyst 2: 0.48
Package 1, sample 2, analyst 1: 0.48
Package 1, sample 2, analyst 2: 0.48
Package 2, sample 1, analyst 1: 0.84

[a] Only leverages for the nonrobust analysis are shown. Those from the robust analysis were similar.

residual. The SAMPLE residuals for this package number will all be larger than the corresponding MEASUREMENT residuals. This is because the partitioning between the two is entirely determined by the relative sizes of the SAMPLE and MEASUREMENT variances.

This difference is also reflected in the SAMPLE and MEASUREMENT leverages for package 2. Recalling that the nominal standard error of a residual is $[\sigma^2\{1 - \text{lev}\}]^{1/2}$, it is seen that the ratio of SAMPLE residual to SAMPLE standard deviation will always be larger than the corresponding ratio of MEASUREMENT residual to MEASUREMENT standard deviation. From equations (8) and (9), one sees that when the single observation from package 2 is an outlier, it will be treated more as a SAMPLE outlier than as a MEASUREMENT outlier.

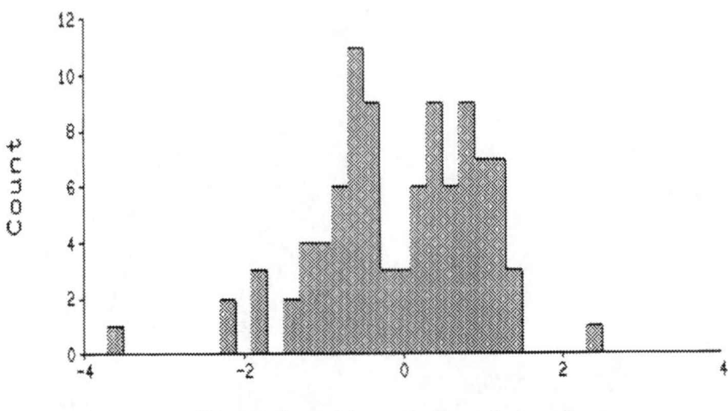

FIGURE 5 Bulk product physical property: histogram of standardized LOT residuals, obtained by Fellner's robust method.

Because the PACKAGE and ANALYST leverages are so small, only the residuals associated with the other three variance components will be considered further. Standardized residuals are used. The standardized residual is the residual divided by its nominal standard error.

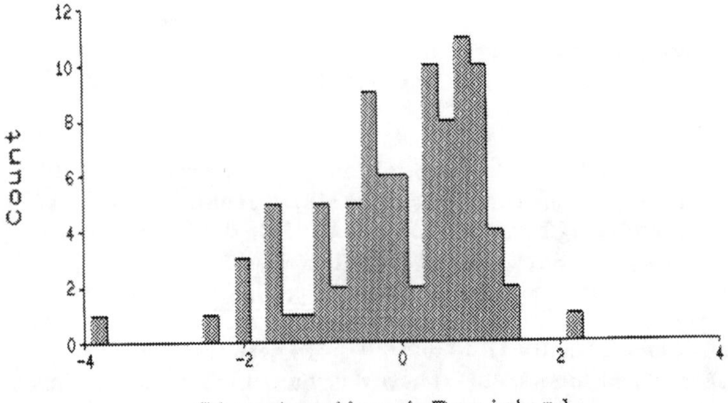

FIGURE 6 Bulk product physical property: histogram of standardized LOT residuals, obtained by restricted maximum likelihood.

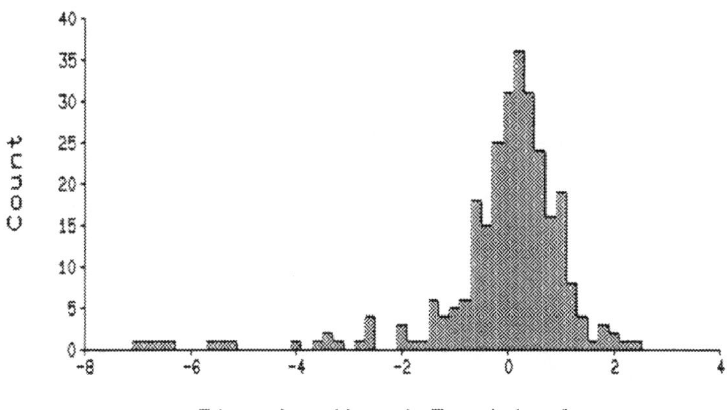

FIGURE 7 Bulk product physical property: histogram of standardized SAMPLE residuals, obtained by Fellner's robust method.

Figures 5 and 6 give histograms of the LOT standardized residuals, as obtained respectively from the robust and nonrobust analyses. The robust analysis gives a decidedly bimodal distribution. There is also a moderate low outlier. The histogram from the nonrobust analysis suggests the bimodality as well, but the picture is muddied because of the skewing effect of the low outliers in the data.

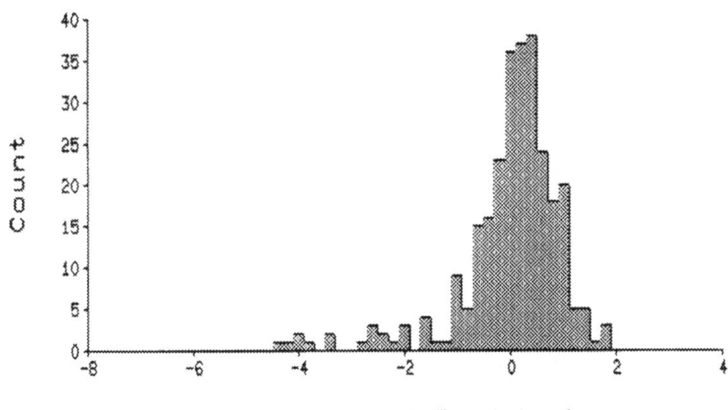

FIGURE 8 Bulk product physical property: histogram of standardized SAMPLE residuals, obtained by restricted maximum likelihood.

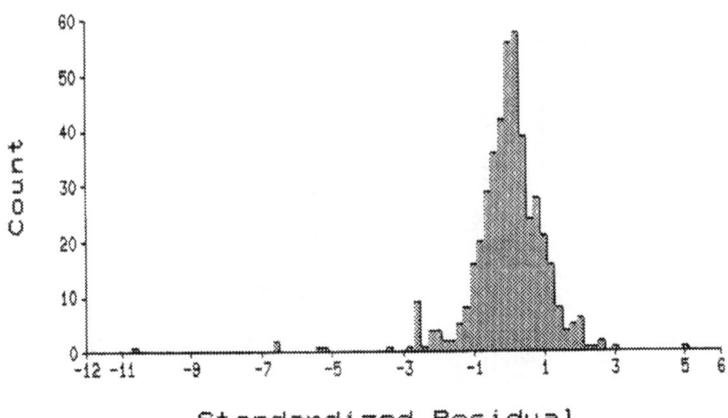

FIGURE 9 Bulk product physical property: histogram of standardized MEASUREMENT residuals, obtained by Fellner's robust method.

Figures 7 and 8 give histograms of the SAMPLE standardized residuals from the robust and nonrobust analyses, respectively. The two distributions are similar except for the low outlying samples, which show up much more clearly in the robust histogram. A similar statement can be made about Figures 9 and 10, which give the robust and nonrobust histograms of the MEASUREMENT standardized residuals.

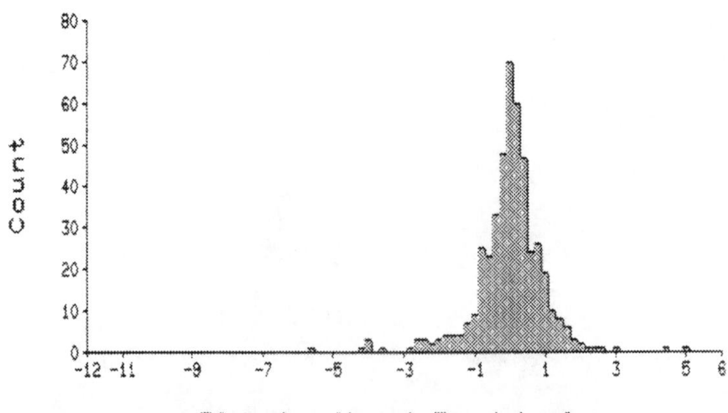

FIGURE 10 Bulk product physical property: histogram of standardized MEASUREMENT residuals, obtained by restricted maximum likelihood.

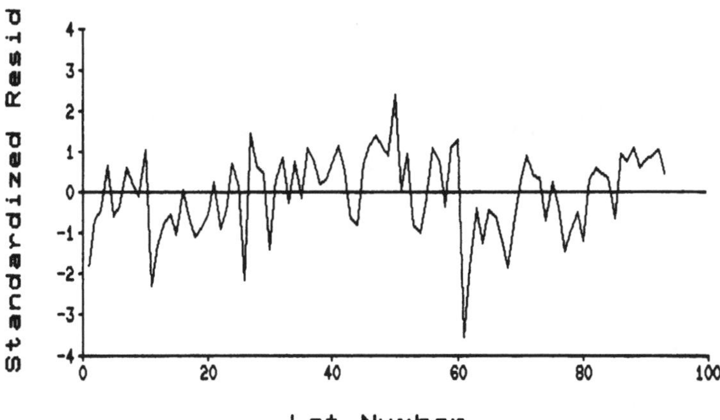

FIGURE 11 Bulk product physical property: standardized LOT residuals, obtained by Fellner's robust method, plotted by lot number.

Figures 11 and 12 give sequence plots of the LOT standardized residuals, again for the robust and nonrobust analyses. Both plots make it clear that the bimodality noted in Figures 5 and 6 is due to shifts in the process over time. The moderate low outlier is seen to be lot 61. As

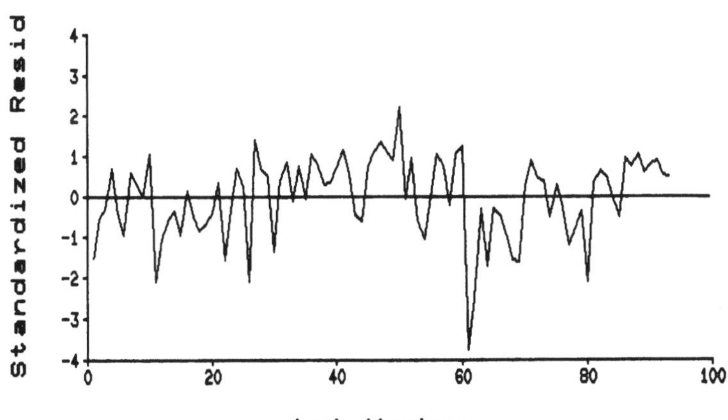

FIGURE 12 Bulk product physical property: standardized LOT residuals, obtained by restricted maximum likelihood, plotted by lot number.

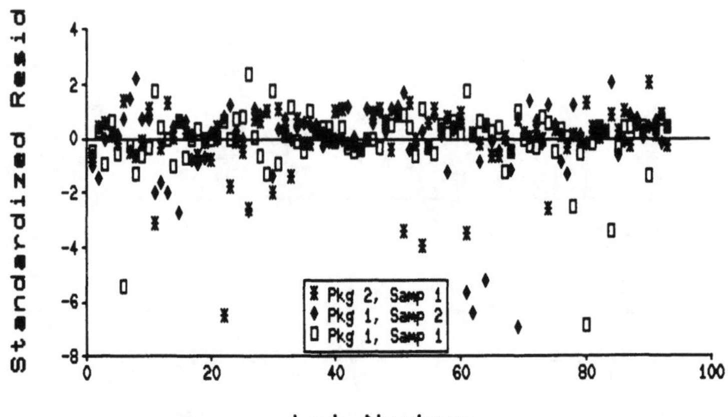

FIGURE 13 Bulk product physical property: standardized SAMPLE residuals, obtained by Fellner's robust method, plotted by lot number.

noted earlier, two of the three samples taken from lot 61 were low outliers.

Figures 13 and 14 give plots of the SAMPLE standardized residuals by lot number, labeled by package and sample. There is a tendency for the outliers to cluster with respect to time. There were no outliers

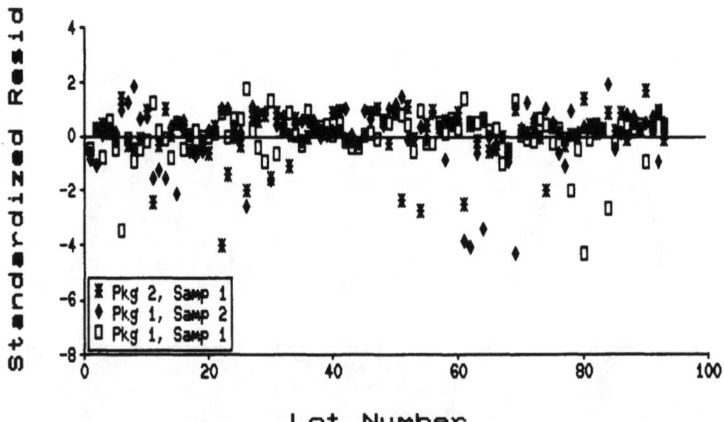

FIGURE 14 Bulk product physical property: standardized SAMPLE residuals, obtained by restricted maximum likelihood, plotted by lot number.

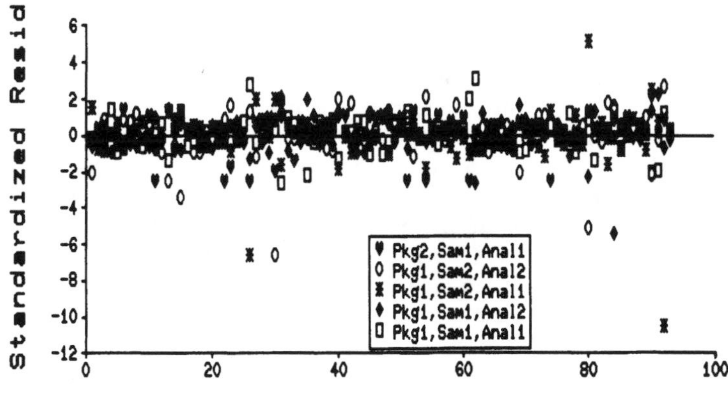

FIGURE 15 Bulk product physical property: standardized MEASUREMENT residuals, obtained by Fellner's robust method, plotted by lot number.

between lots 35 and 50. Moreover, there is a tendency for the outliers to cluster with respect to label. For example, from lots 61 through 69 the low samples are mostly from package 1, sample 2. From lots 78 through 90, the low samples are mostly from package 1, sample 1. This would suggest that the sampling biases, which caused the differences in the five means in Table 3, were not constant over time.

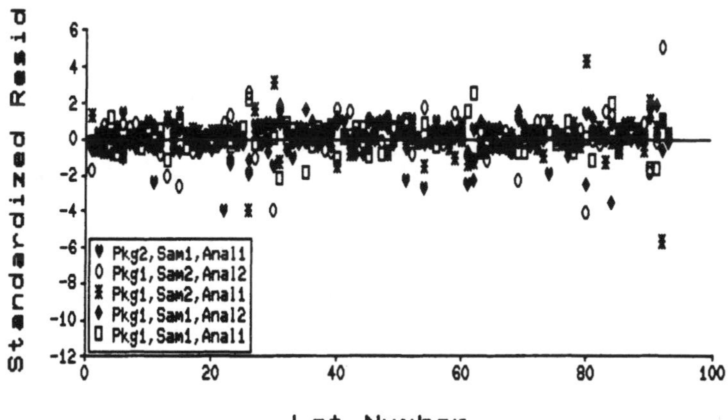

FIGURE 16 Bulk product physical property: standardized MEASUREMENT residuals, obtained by restricted maximum likelihood, plotted by lot number.

Figures 15 and 16 give plots of the MEASUREMENT standardized residuals by lot number, labeled by package, sample, and analyst. The robust plot shows the outliers more clearly; however, there are fewer of them! The explanation for this is enlightening.

For example, in lot 22, the standardized residual for package 2, sample 1, analyst 1 appears an outlier in the nonrobust plot but not in the robust plot. The data value corresponding to this residual is 11.75. This value is either an outlying measurement or an outlying sample. The robust analysis makes the latter "choice" because the SAMPLE variance is much larger than the MEASUREMENT variance. As a result, the SAMPLE residual is made larger and the MEASUREMENT residual smaller.

In lot 92, the standardized MEASUREMENT residuals for package 1, sample 2, analysts 1 and 2 are plotted as outliers, one high and one low in the nonrobust plot. In the robust plot, only the residual for analyst 1 appears to be an outlier. The data show a single low outlier, 16.79, for this lot. The robust analysis has handled this appropriately, while the nonrobust analysis has, in effect, "split the difference" between the two analysts.

At first glance, lot 80 seems to be similar to lot 92. Again, package 1, sample 2 received two discordant measurements. However, in this case, package 1, sample 1 is a low outlier, and there is no clear indication of what the true mean for this lot should be. Thus, both the robust and nonrobust analyses split the difference between the two analysts.

7. CONCLUSION

Henderson's method maximizes the diagnostic information for the residuals associated with the largest variance components. By contrast, the method of unweighted means maximizes the diagnostic information for the residuals highest in the nesting hierarchy. The latter has no obvious rationalization.

One way of thinking about robust analysis is that, rather than tending to minimize the sizes of outlying residuals, it tends to minimize their number. This gives the dual advantage of making the outliers more visible and making everything else less affected by the outliers. Thus parameter estimates reflect the body of the data, and the nonoutlying residuals are better able to display other departures from the assumed model.

REFERENCES

Andrews, D. F. (1979). The robustness of residual displays. Robustness in Statistics, R. L. Launer and G. N. Wilkinson, eds., pp. 12-32. Academic Press, New York.

Bickel, P. J. (1976). Another look at robustness: A review of reviews and some new developments. Scand. J. Stat. 3: 145-168.

Fellner, W. H. (1986). Robust estimation of variance components. Technometrics 28: 51-60.

Fellner, W. H. (1987). Sparse matrices, and the estimation of variance components by likelihood methods. Commun. Stat. B 16: 439-463.

Harville, D. A. (1977). Maximum likelihood approaches to variance component estimation and to related problems. J. Am. Stat. Assoc. 72: 320-340.

Henderson, C. R. (1963). Selection index and expected genetic advance. Statistical Genetics and Plant Breeding. National Research Council Publ. 982, pp. 141-163. National Academy of Sciences, Washington, D.C.

Huber, P. J. (1964). Robust estimation of a location parameter. Ann. Math. Stat. 35: 73-101.

Patterson, H. D., and Thompson, R. (1971). Recovery of interblock information when block sizes are unequal. Biometrika 58: 545-554.

Rocke, D. M. (1983). Robust statistical analysis of interlaboratory studies. Biometrika 70: 421-431.

Searle, S. R. (1971). Linear Models. Wiley, New York.

Shoemaker, L. H. (1980). Robust Estimates and Tests for the One-Sample Scale Model, with Applications to Variance Component Models. University Microfilms International, Ann Arbor, Mich.

Snee, R. D. (1983). Graphical analysis of process variation studies. J. Qual. Technol. 15: 76-88.

Thompson, W. A., Jr. (1962). The problem of negative estimates of variance components. Ann. Math. Stat. 33: 273-289.

Yates, F. (1934). The analysis of multiple classifications with unequal numbers in the different classes. J. Am. Stat. Assoc. 29: 51-66.

Part III

Forecasting and Robust Regression

8

Robust Time Series Analysis—An L_1 Approach

KENNETH O. COGGER School of Business, The University of Kansas, Lawrence, Kansas

1. INTRODUCTION

Robust estimation of time series models has been a subject of limited investigation in the past five years. As in the more extensively researched multiple regression case, one of the most robust, and computationally simple, robust procedures for time series analysis is to utilize an L_1 norm and minimize absolute error rather than squared error. Statistical properties of resulting parameter estimates are becoming better known, but these results are widely scattered in the literature. The present work summarizes some of the known properties of L_1 estimates and discusses how existing computer algorithms can be modified to permit direct estimation of prediction intervals. An actual time series analysis is described and comparisons are made with results based on the usual least squares approach to the same data. Appropriate statistical tests are conducted.

2. COMPUTATIONAL PROCEDURES

In the context of multiple regression, the use of a least squares criterion for the estimation of parameters can be justified in several ways. One could invoke the celebrated Gauss-Markov theorem and argue that least squares estimators have minimum variance among all linear estimators provided that the assumption of uncorrelated errors is met. An alternative justification might be that least squares estimators are maximum likelihood estimators if the assumption of normal error terms is merited. From a computational viewpoint, least squares procedures are easily programmed and computer algorithms are widely available. By the same line of reasoning, one would be tempted to consider alternatives to least squares when its underlying assumptions are uncertain, provided that the additional computational burden is not severe.

One alternative to least squares estimation is to uitilize an absolute error criterion. If error terms are not normally distributed, the Gauss-Markov result does not apply even if the errors are statistically independent, since L_1 estimators are not linear in the observations. In such situations, alternatives to least squares may be justified. Unfortunately, computationally efficient algorithms for L_1 estimation were not available until about 10 years ago. Charnes et al. (1955) were the first to show that such estimates could be obtained through the use of linear programming, a mathematical programming procedure which had, at that time, been only recently developed. One of the most efficient algorithms available today is the one described in Barrodale and Roberts (1973), which takes full advantage of the special character of the linear program when used in a regression context. With this algorithm, problems with 100 observations and several independent variables may be solved on a microcomputer in a few minutes, even with interpreted BASIC programs. Considerably shorter times are possible with compiled programs utilizing a numeric coprocessor chip.

With the disappearance of computational difficulties, the only remaining drawback to the use of L_1 estimators has been, until recently, the lack of understanding of their statistical properties. It has long been understood that they are robust estimators, in exactly the same sense that the special case of the sample median is more robust than the sample average. But what has been needed is some indication of whether and under what conditions such estimators are unbiased, of their sampling variance, and of the nature of their sampling distribution. The first definitive answer, in the case of multiple regression, came from Koenker and Bassett (1978). They showed that under very weak conditions, L_1 regression estimators were consistent and asymptotically normal with covariance matrix

$$V = \frac{(X'X)^{-1}}{4*f^2} \tag{1}$$

where X is the regression design matrix and f is the value of the error density at the origin. For normally distributed errors, variances given by equation (1) are about 57% larger than those of ordinary least squares. However, equation (1) also reveals that L_1 estimates have lower variance than those of least squares whenever the error variance exceeds $1/4*f^2$.

The paper by Koenker and Bassett also extends the idea of minimizing absolute error to that of minimizing weighted absolute error, allowing different weights to be assigned to positive and negative errors. They thus introduced the concept of regression quantiles, which we will apply in the next section of this chapter.

Many of the concepts and results discussed above extend directly to autoregressive time series analysis. Here, the independent variables are simply lagged values of the dependent variable. Similar extensions are not easily made to moving-average time series models. With such models, the mathematical programming problem becomes nonlinear and no special-purpose algorithms are yet available. General-purpose algorithms are inefficient for the special structure of moving-average estimation problems, although they certainly could be employed if desired. An alternative would be to continue to use the highly efficient Barrodale and Roberts algorithm and approximate moving-average models with very high-order autoregressions, followed by an application of the Durbin (1959) procedure to determine the moving-average coefficients. For simplicity of presentation, this chapter restricts attention to the case of pure autoregressive models.

The asymptotic results for L_1 estimators in the case of the multiple regression model do, in many cases, extend to the case of autoregressive models, provided that $(X'X)^{-1}$ in equation (1) is replaced by its probability limit, analogous to the case of least squares. This results in L1 estimates of autoregressive parameters having an asymptotic normal distribution identical to that in the ordinary regression case with the exception that the covariance matrix is

$$V = \text{plim} \frac{(X'X)^{-1}}{4*f^2} \tag{2}$$

This result, as noted by Gross and Steiger (1979), appears to hold under very mild conditions but it is not yet known whether it holds in the case of distributions with infinite variance, as it does in the case of ordinary regression. These authors do, however, prove that L_1 autoregression estimates are strongly consistent even with distributions characterized by infinite variance, a remarkable property. Further results on the convergence of these estimates appear in An and Chen (1982). Very few results are available for small-sample properties, although Gross and Steiger (1979) provide limited Monte Carlo sampling results.

3. A COMPARATIVE TIME SERIES ANALYSIS

The time series chosen for a representative analysis is the Personal Saving Rate, recorded on a quarterly basis from 1955 to 1984. Taken from the Business Conditions Digest, this time series is seasonally adjusted prior to publication.

This chapter is not concerned with robust procedures for the model identification stage of time series analysis, only with the parameter estimation stage. Thus the identification of the type of model utilizes the standard examination of sample autocorrelations and partial autocorrelations. These, of course, are not robust statistics. I have found, however, that even when they are misleading due to outliers and so forth, appropriate diagnostics on the robustly estimated model will provide sufficient feedback.

With our sample size of $T = 120$, partial autocorrelations exceeding 0.183 in absolute value are significant at the $p = .05$ level. Quite clearly, Table 1 gives strong evidence of a first-order autoregressive

TABLE 1

Lag autocorrelation	Autocorrelation	Partial
1	0.769	0.769
2	0.619	0.066
3	0.445	-0.124
4	0.320	-0.008
5	0.225	0.007
6	0.141	-0.043
7	0.077	-0.021
8	0.100	0.170
9	0.105	0.007
10	0.133	0.032

TABLE 2 Results Based on Maximum Likelihood (Least Squares)

Coefficient	Estimate	Standard deviation	T ratio
AR 1	0.7781	0.0580	13.42
Constant	1.3585	0.0636	21.35

Mean squared error = 0.4853

Forecast from period 120

Forecast = 6.0268
Lower 2.5% = 4.6611
Upper 97.5% = 7.3926

model, and this model is selected for parameter estimation and subsequent diagnostic checking. Table 2 gives results for the usual maximum likelihood estimates based on a normality assumption. With large sample sizes, as in the present case, this is nearly equivalent to least squares. Table 3 provides corresponding results for L_1 estimates. Programs used were MINITAB and a BASIC program using the Barrodale and Roberts algorithm. Both were run on a PC-XT compatible.

Two important aspects of Table 3 must be mentioned. First, the calculation of the standard deviations requires the estimation of f, the error density at the origin. See equation (2). Consistent estimates of f may be calculated in many ways. This chapter uses the ordered regression

TABLE 3 Results Based on L_1 Estimation (Least Absolute Error)

Coefficient	Estimate	Standard deviation	T-ratio
AR 1	0.7586	0.0517	14.67
Constant	1.4345	0.0567	25.30

Average absolute error = 0.5141

Forecast from period 120

Forecast = 5.9862
Lower 2.5% = 4.5618
Upper 97.5% = 7.2265

residuals. A plot of the ordered residuals against their order indices yielded an approximately linear relationship near the median residual. The slope of this line is equivalent to an average of several of the estimates described by Cox and Hinkley (1974, p. 470). We employed ordered residuals 49 to 64 to estimate the slope, from which the estimate f = 0.806 was obtained from the formula

$$1/f = T * \text{slope} \tag{3}$$

Given the many alternatives to (3) for the estimation of densities, much future work on this aspect of the estimation problem seems warranted.

The second important aspect of Table 3 is the calculation of forecast limits. These were calculated directly, using the notion of regression quantiles, by estimating coefficients with differential weights on positive and negative residuals. For the upper limit, positive residuals were weighted by a factor of 0.975 and negative residuals were weighted by 0.025. For the lower forecast limit, these weights were reversed. The resulting autoregressive coefficient estimates were then combined with the actual figure for period 120 to arrive at forecast limits for period 121. One convenient aspect of this procedure is that the distribution of the errors need not be assumed normal, as is the case in Table 2, and the empirical distribution of the residuals need not be examined directly.

3. CONCLUSION

We have presented a side-by-side analysis of a time series using a standard and a robust approach, describing how tests of significance may be performed with the latter and illustrating how confidence limits on forecasts may be empirically calculated. One of the most important uses of a robust analysis is as an insurance policy. If, as in the current case study, the results of a standard approach and those of a robust approach are similar, this should give greater confidence in the former. On the other hand, if the results are in some conflict, it may be the case that standard statistical assumptions are faulty, causing the robust analysis to be given greater credence, particularly when prediction intervals are desired. The current availability of efficient computational procedures as well as an increasing awareness of statistical properties should encourage more frequent use of such robust time series analyses.

REFERENCES

An, H., and Chen, Z. (1982). On convergence of LAD estimates in autoregression with infinite variance. J. Multivariate Anal. 12: 335-345.

Barrodale, I., and Roberts, F. D. K. (1973). An improved algorithm for discrete L1 approximation. SIAM J. Numer. Anal. 10: 839-848.

Charnes, A., Cooper, W. W., and Ferguson, R. O. (1955). Optimal estimation of executive compensation by linear programming. Manage. Sci. 1: 138-151.

Cox, D. R., and Hinkley, D. V. (1974). Theoretical Statistics. Chapman & Hall, London.

Durbin, J. (1959). Efficient estimation of parameters in moving average models. Biometrika 46: 306-316.

Gross, S., and Steiger, W. L. (1979). Least absolute deviation estimates in autoregression with infinite variance. J. Appl. Probab. 16: 104-116.

Koenker, R. W., and Bassett, G. W. (1978). Regression quantiles. Econometrica 46: 33-50.

9

Robust Regression and Input-Output Forecasts

SHEILA M. LAWRENCE Rutgers University, North Brunswick, New Jersey

ELLIE HAKAK ELEONOMIC, Flushing, New York

M. GOLD New York University, New York, New York

1. NATIONAL INPUT-OUTPUT MODELLING

1.1 The Structure of National Static Open Models

Input-output models are analytical frameworks incorporating detailed information about applied production technologies, interindustry transactions, and final purchases. First developed and implemented by Wassily Leontief (1941) as a system for the empirical study of national economies, the analytical form has been developed by many users into a family of frameworks addressing a broad range of tasks. (A review of basic methods and various applications can be found in, e.g., Leontief, 1966.) These systems are popular forecasting and analytical tools.

By maintaining explicit, disaggregated accounts about sectoral interaction, input-output systems are capable of modelling the diverse indirect sectoral "feedback" requirements needed to meet direct demands for the products of particular industries. The best known and most widely applied input-output model is the static open system—the most basic of the various input-output frameworks from both empirical and theoretical perspectives. From this basis, more complicated models have been developed, including dynamic, regional, and multiregional systems (see, e.g., Leontief et al., 1953).

The foundation of all input-output systems is a basic accounting relation. Total output less output used in intermediary production leaves output available for final use. In the case of empirically developed systems, a set of mutually exclusive industrial sectors is defined and indexed. Criteria for categorizing activities are complicated but the basic goal is to group together activities which generate similar products through the use of techniques including similar input requirements.

Once a comprehensive set of sectors has been defined, the annual purchases of each sector from each sector are arrayed in columns. A companion set of accounts is also kept of each sector's "value added" payments to "primary factors"—wages, profits, rents, and so forth. Consider sector i. Sector j payments to sector i will be denoted as h_{ij}. Sector j wages will be denoted as w_j, profits as p_j, and rents as r_j. A set of transaction input accounts for sector j would appear as

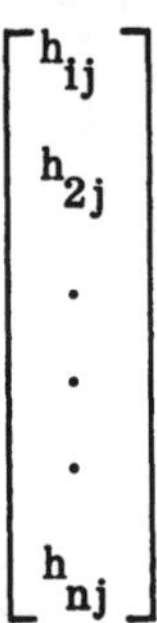

$$\begin{bmatrix} w_j \\ p_j \\ r_j \end{bmatrix}$$

where h_{ij} is the account of sector 1 shipments to j and so on. The array above details the direct input requirements of sector j for various commodities (1, 2, . . . , n) and primary factors over the course of a year. Consider what happens when column arrays are joined together so that row indices and column indices correspond:

$$\begin{bmatrix} h_{11} & \cdots & h_{1j} & \cdots & h_{1n} \\ \cdot & & \cdot & & \cdot \\ \cdot & & \cdot & & \cdot \\ \cdot & & \cdot & & \cdot \\ h_{i1} & \cdots & h_{ij} & \cdots & h_{in} \\ \cdot & & \cdot & & \cdot \\ \cdot & & \cdot & & \cdot \\ \cdot & & \cdot & & \cdot \\ h_{n1} & \cdots & h_{nj} & \cdots & h_{nn} \end{bmatrix}$$

$$\begin{bmatrix} w_1 & \cdots & w_j & \cdots & w_n \\ p_1 & \cdots & p_j & \cdots & p_n \\ r_1 & \cdots & r_j & \cdots & r_n \end{bmatrix}$$

The result is an array set in which the rows detail the flow of output from each sector to its various sectoral customers. By appending to this array additional information concerning final use of sectoral output, that is, a sectoral disaggregation of GNP, the entire activity of an economy may be completely charted, in sectoral detail, for the given year.

To create an analytical model from the set of transaction accounts above, each column of the transaction matrix is divided by the value of total output of the corresponding column index sector (i.e., each element h_{ij} is divided by the total output of sector j, x_j). The resulting array is commonly known as a technology matrix or matrix of technical coefficients and is denoted by the letter A. Each element a_{ij} of the matrix A denotes the sector j direct requirement for sector i goods per dollar of sector j output.

The system of accounts is now brought together as a model by introducing a basic relationship that production supply must equal intermediate and final consumption demand or

$$X_i = \sum_j a_{ij}X_j + y_i$$

for all i. In this equation, X_i represents the total annual output of sector i, and y_i represents the value of final use of sector i goods, that is, "final demand." The balance relation of the entire system may be portrayed in matrix form as

$$\begin{bmatrix} X_1 \\ \\ \\ X_i \\ \\ \\ X_n \end{bmatrix} - \begin{bmatrix} a_{11} & \cdots & a_{1j} & \cdots & a_{1n} \\ \\ \\ a_{i1} & \cdots & a_{ij} & \cdots & a_{in} \\ \\ \\ a_{n1} & \cdots & a_{nj} & \cdots & a_{nn} \end{bmatrix} \begin{bmatrix} X_1 \\ \\ \\ X_i \\ \\ \\ X_n \end{bmatrix} = \begin{bmatrix} y_1 \\ \\ \\ y_i \\ \\ \\ y_n \end{bmatrix}$$

which may be concisely written as

$$x - Ax = y \quad \text{or} \quad (I - A)x = y$$

where x is a vector of total outputs, y is a vector of final demands, A is the technology matrix, and I is the identity matrix with units on the diagonal and zeros elsewhere.

The matrix equation above is a basic representation of a static input-output system. It is called an "open" system because the vector of final demands above is not endogenously restricted. Once the coefficients of the A matrix have been established, input-output can be used to project total output requirements necessary to meet a given set of final demand targets.

Note the difference between "total requirements to meet final demand" and "direct interindustry requirements to meet a level of production." Direct interindustry requirements associated with a vector of production levels x are the vector Ax. But the requirements to meet those requirements (e.g., the inputs of coal, iron, and scrap needed to meet the steel sectoral requirement in Ax) are to be found in the multiplication AAx or A^2x and so on.

Total requirements may be found by solving the initial equation system for the vector x, that is,

$$(I - A)^{-1}y = x$$

The matrix $(I - A)^{-1}$ is often called the Leontief inverse. Denote this matrix by the letter M. The elements of this matrix, m_{ij}, denote the total direct and indirect requirements of commodity i to meet one unit of final demand for sector j goods. Thus:

$$x_i = \sum_j m_{ij}y_j$$

for each sector i.

The interpretation of Leontief inverse elements as being total direct and indirect requirements multipliers may be supported by noting that since (in a viable economy with some value-added payments in each industry) individual elements and also row sums are both restricted to a range from zero to just under 1, the Leontief inverse is the limit of the converging infinite series $(I + A + A^2 + A^3 \cdots)$. In the context of input-output, the series may be interpreted as saying that total requirements to meet final demand represent a sum: final demand, plus the requirements to meet the final demand, plus the requirements to meet those requirements, etc. By charting sectoral interactions in the framework of a simultaneous equation system, input-output effectively captures in extensive detail the direct and indirect output activity levels needed to meet a set of final demands.

1.2 Components of Final Demand and the Volatility of Aggregate Measures

National input-output tables for the U.S. economy are regularly prepared by the Department of Commerce for quintennial census years. The Department of Commerce publishes a detailed set of accounts with approximately 500 sectors and a more highly aggregated 85-sector model. Release of the tables lags five to seven years. Tables for the 1977 U.S. economy at the 85-sector level of aggregation were published in the <u>Survey of Current Business</u> in May 1984. Tables for the 1982 economy are expected shortly.

TABLE 1 Input-Output Transactions Totals for the U.S. Economy (Figures in Millions of Dollars at Producers Prices)

		Total intermediate use	Personal consumption expenditures	Gross private fixed investment	Change in business inventories	Exports	Imports	Federal Government purchases			State and local government purchases			Total final demand	Total commodity output	Commodity number
								Total	National defense	Nondefense	Total	Education	Other			
			91	92	93	94	95		96	97		98	99			
1	Livestock and livestock products	47,384	2,511		-1,183	199	-360	6	1	5	48	24	24	1,219	48,603	1
2	Other agricultural products	38,279	7,726		1,832	12,523	-1,047	3,496		3,496	367	191	177	24,897	63,176	2
3	Forestry and fishery products	6,346	788		34	214	-1,302	-828		-828	-81	4	-85	-1,175	5,170	3
4	Agricultural, forestry, and fishery services	8,095	353			24	-3	61	4	57	392	165	227	828	8,923	4
5	Iron and ferroalloy ores mining	3,548			-437	326	-1,173	-49	-49					-1,335	2,213	5
6	Nonferrous metal ores mining	3,257		374	57	203	-728	-16	-16	(*)				-110	3,147	6
7	Coal mining	14,121	215		161	2,096	-86	31	22	9	109	49	60	2,525	16,646	7
8	Crude petroleum and natural gas	77,477		116	690	202	-35,062	100	1	99				-33,965	43,523	8
9	Stone and clay mining and quarrying	4,930	20		79	181	-286	-2	-2		-39		-39	-47	4,883	9
10	Chemical and fertilizer mineral mining	1,298	2		39	228	-231	3		3	87		87	128	1,426	10
11	New construction			150,890		1		7,450	2,361	5,089	32,354	5,585	26,769	190,694	190,694	11
12	Maintenance and repair construction	57,525				26		3,350	2,265	1,086	12,739	4,351	8,388	16,115	73,640	12
13	Ordnance and accessories	621	630	22	115	1,530	-99	5,978	5,157	821	45		45	8,220	8,841	13
14	Food and kindred products	75,195	113,507		1,617	7,308	-8,358	604	161	443	2,983	2,045	938	117,660	192,855	14
15	Tobacco manufactures	2,628	8,437		365	1,664	-272							10,195	12,823	15
16	Broad and narrow fabrics, yarn and thread mills	24,338	882		1,082	1,148	-1,075	55	50	5	59	27	32	2,152	26,489	16
17	Miscellaneous textile goods and floor coverings	5,948	2,045	892	187	342	-402	16	2	14	21	4	17	3,101	9,049	17
18	Apparel	10,305	33,194		2,472	733	-5,865	344	344		288	7	281	31,167	41,472	18
19	Miscellaneous fabricated textile products	4,675	4,068		222	332	-255	73	54	20	187	35	152	4,627	9,303	19
20	Lumber and wood products, except containers	38,243	548	11	1,329	1,928	-3,537	23	18	5	72	49	23	373	38,616	20
21	Wood containers	525			8	10	-40	5	3	2				-17	508	21
22	Household furniture	571	8,642	725	360	203	-475	54	9	45	56	40	16	9,566	10,137	22
23	Other furniture and fixtures	830	566	4,325	131	91	-269	105	23	82	664	411	253	5,612	6,442	23
24	Paper and allied products, except containers	31,919	5,307		739	2,150	-3,725	128	32	96	1,180	525	654	5,779	37,698	24
25	Paperboard containers and boxes	12,359	192		181	178	-13	43	26	17	92	40	52	673	13,033	25
26	Printing and publishing	16,718	10,237		596	702	-360	332	138	195	3,624	2,117	1,507	15,131	31,849	26
27	Chemicals and selected chemical products	56,407	1,149	541	1,083	6,273	-4,370	1,275	1,071	204	906	330	576	6,857	63,263	27
28	Plastics and synthetic materials	20,775			259	1,734	-495	37	34	3	2	2	(*)	1,538	22,313	28
29	Drugs, cleaning and toilet preparations	8,541	16,921		600	1,703	-1,338	330	198	132	1,970	239	1,731	20,184	28,725	29
30	Paints and allied products	5,600	168		148	162	-8	3	(*)	3	119	102	17	594	6,194	30
31	Petroleum refining and related industries	57,315	38,595		3,046	2,693	-11,366	1,875	2,043	-169	3,956	1,795	2,160	38,799	96,114	31
32	Rubber and miscellaneous plastics products	32,019	6,444	58	1,366	1,532	-2,527	309	213	96	590	137	453	7,772	39,791	32

33	Leather tanning and finishing	1,531			18	166	-175	1	(*)	1				10	1,541	33
34	Footwear and other leather products	575	7,610		209	144	-2,493	24	17	7	28		28	5,522	6,097	34
35	Glass and glass products	7,909	829		163	503	-466	16	7	9	282	87	195	1,327	9,236	35
36	Stone and clay products	24,239	1,123		717	604	-1,247	66	22	44	87	36	51	1,349	25,589	36
37	Primary iron and steel manufacturing	67,833	11	5	1,274	1,580	-7,256	157	119	38	21	4	16	-4,209	63,623	37
38	Primary nonferrous metals manufacturing	42,579	48	106	1,007	1,512	-4,747	178	88	90	18	1	17	-1,878	40,702	38
39	Metal containers	8,256		29	164	79	-55	54	54		25	24	1	296	8,551	39
40	Heating, plumbing, and structural metal products	20,918	374	3,055	902	1,126	-251	890	634	256				6,096	27,014	40
41	Screw machine products and stampings	18,120	798		368	1,059	-618	104	72	32	161	124	38	1,871	19,992	41
42	Other fabricated metal products	23,522	2,038	1,591	830	1,466	-1,757	408	288	120	174	95	78	4,751	28,273	42
43	Engines and turbines	6,184	207	1,663	423	1,993	-468	751	725	25	112		112	4,681	10,865	43
44	Farm and garden machinery	1,956	106	8,410	661	1,240	-1,057	21	17	4	80	20	60	9,460	11,416	44
45	Construction and mining machinery	3,596		8,692	603	4,421	-814	156	127	29	329		329	13,386	16,982	45
46	Materials handling machinery and equipment	1,546		2,984	104	427	-201	113	76	36	3	1	1	3,430	4,976	46
47	Metalworking machinery and equipment	5,267	281	7,507	400	1,087	-953	198	116	82	59	39	20	8,580	13,846	47
48	Special industry machinery and equipment	2,074	92	5,209	227	2,354	-1,248	84	64	20	28	26	1	6,745	8,818	48
49	General industrial machinery and equipment	9,451		5,080	474	2,214	-994	283	193	90	48		48	7,105	16,556	49
50	Miscellaneous machinery, except electrical	8,038	40	29	184	157	-92	102	36	66	34	17	17	454	8,492	50
51	Office, computing, and accounting machines	3,873	420	7,432	558	3,476	-1,550	1,217	867	351	373	236	138	11,826	15,798	51
52	Service industry machines	6,264	432	2,986	317	1,156	-133	89	54	36	276	217	59	5,122	11,386	52
53	Electric industrial equipment and apparatus	10,298	91	5,854	586	2,072	-1,488	795	552	242	114	48	67	8,024	18,321	53
54	Household appliances	1,775	7,014	1,607	174	657	-965	33	28	5	78	32	46	8,597	10,371	54
55	Electric lighting and wiring equipment	5,964	1,318	97	361	460	-240	67	47	21	171	128	43	2,234	8,199	55
56	Radio, TV, and communication equipment	6,836	8,328	10,620	703	2,498	-5,716	4,794	4,395	400	269	183	86	21,497	28,333	56
57	Electronic components and accessories	12,720	529	35	490	2,468	-2,226	715	454	261	60	20	40	2,069	14,790	57
58	Miscellaneous electrical machinery and supplies	4,335	2,003	1,491	383	859	-763	164	83	81	141	19	122	4,277	8,612	58
59	Motor vehicles and equipment	39,822	46,124	30,854	4,368	10,963	-18,253	976	685	291	2,050	651	1,399	77,084	116,906	59
60	Aircraft and parts	5,597	427	2,777	186	7,159	-760	9,795	9,166	629	8		8	19,592	25,189	60
61	Other transportation equipment	2,465	7,068	8,328	597	975	-1,284	2,997	2,846	151	247	29	218	18,917	21,382	61
62	Scientific and controlling instruments	4,395	1,927	4,570	484	1,976	-1,395	1,107	708	398	567	60	508	9,236	13,631	62
63	Optical, ophthalmic, and photographic equipment	3,687	2,379	4,188	177	1,510	-1,713	679	276	404	792	380	412	8,013	11,700	63
64	Miscellaneous manufacturing	5,775	12,684	1,283	913	1,295	-3,833	83	58	25	836	518	318	13,263	19,038	64
65	Transportation and warehousing	75,440	33,210	1,976	1,020	9,756	-332	3,315	2,728	587	3,879	2,373	1,506	52,823	128,264	65
66	Communications, except radio and TV	23,404	22,394	3,365		985		1,063	502	562	1,636	861	785	29,464	52,868	66
67	Radio and TV broadcasting	182	344											344	526	67
68	Electric, gas, water, and sanitary services	75,722	41,824			276	-2,200	1,524	862	662	5,311	2,466	2,845	46,734	122,456	68
69	Wholesale and retail trade	112,682	222,550	24,668	2,980	12,416	5,376	2,125	1,584	542	3,374	1,254	2,121	273,489	386,171	69
70	Finance and insurance	58,752	65,533		(*)	630	-524	613	7	606	3,574	184	3,390	69,826	128,578	70
71	Real estate and rental	79,438	181,314	10,747		3,705		700	236	464	3,338	446	2,892	199,805	279,243	71
72	Hotels: personal and repair services (exc. auto)	10,924	33,938			29		557	380	177	681	-294	975	35,205	46,129	72
73	Business services	131,330	13,863			3,481	-100	7,053	2,666	4,387	6,343	2,324	4,019	30,640	161,969	73
74	Eating and drinking places	22,153	67,477			81		194	129	66	2,067	-2,762	695	65,685	87,839	74
75	Automobile repair and services	17,170	25,437		7	2	-11	81	48	33	694	153	541	26,210	43,380	75
76	Amusements	7,666	16,018		145	444	-37	189	132	57	159	134	25	16,917	24,583	76
77	Health, educ., & social serv. and nonprofit org	7,093	135,932			75		3,838	764	3,074	15,078	-757	15,835	154,923	162,016	77
78	Federal Government enterprises	10,702	2,692			193		175	140	35	478	48	431	3,538	14,240	78
79	State and local government enterprises	1,259	3,576			1		43	33	10	83	44	39	3,702	4,961	79
80	Noncomparable imports	13,377	8,727	17	35		-26,610	4,436	3,406	1,030	17	15	2	-13,377		80
81	Scrap, used, and secondhand goods	4,949	5,502	-10,297	-102	1,558	-264	-55	-24	-31	959	212	747	-2,699	2,250	81
82	Government industry							65,523	42,213	23,309	138,411	77,533	60,878	203,934	203,934	82
83	Rest of the world industry		-7,221			40,119	9,117	317	14	303				23,464	23,464	83
84	Household industry		5,930											5,930	5,930	84
85	Inventory valuation adjustment				-18,582									-18,582	-18,582	85
I	Total intermediate inputs															I
VA	Value added														1,976,563	VA
88	Compensation of employees														1,165,555	88
89	Indirect business taxes														165,958	89
90	Property-type income														645,051	90
T	Total industry output		1,246,481	314,926	21,700	182,043	-184,154	143,363	92,825	50,538	252,284	105,492	146,712	1,974,563		T

Source: Survey of Current Business, May 1984, p. 57.

TABLE 2 Change in GNP and Gross Investment (Billions of 1972 Constant Dollars)

	Year											
	1974/75	1975/76	1976/77	1977/78	1978/79	1979/80	1980/81	1981/82	1982/83	1983/84	x	o
Δ$GNP[a]	-14	66	72	69	40	-4	37	-32	55	104		
Δ$GPDI[b]	-40	30	39	23	-1	-27	22	-37	27	69		
ΔDGPFI	-22	15	24	20	8	-16	7	-15	20	40		
%ΔGNP	- 1.2	5.4	5.5	5.0	2.8	-.3	2.5	- 2.1	3.7	6.8	2.81	3.08
%ΔGPDI	-20.8	19.2	16.1	10.5	-.2	-11.8	10.7	-15.8	13.7	31.2	5.78	16.85

[a] Δ denotes "change in."

[b] Gross private domestic investment (GPDI) = gross private fixed investment (GPFI) + change in business inventories.

Source: Statistical Abstract of the United States, 1986, p. 432.

Table 1 presents total intermediate sectoral use, detailed sectoral final demand, and total commodity output for the 1955 U.S. economy. Since final demand is a sectoral disaggregation of gross national product (GNP), it may be presented in terms of component parts, i.e., vectors of sectoral consumption, investment, net exports, and government spending. The vectors for these subdivisions are given in columns 91 to 99. Column totals are recorded on the bottom row.

Clearly, the most important component of final demand from the point of view of aggregate scale is consumption, which is nearly two-thirds of total final demand. Combined government purchases account for 20% of 1977 U.S. final demand. Net exports had a negligible impact on the total of final purchases. Gross investment represented the remainder—about 16%.

Gross investment, while not being a particularly prominent component of GNP with respect to scale, is indeed a very important component with respect to GNP volatility. A great deal of the volatility in annual GNP may be associated with investment fluctuations, and the movements in the time series of each are highly correlated. Consequently, accurate forecasts of investment fluctuation will be very useful in maintaining accurate forecasts of GNP movements.

1.3 The Application of Controls and Input-Output Forecasts

As noted above, total sectoral output forecasts may be established by means of the input-output relation

$$(I - A_t)^{-1} y_t = x_t$$

where t is a time subscript. Critical to this forecasting function is an accurate forecast for y_t.

It may be possible to establish short-term forecasts for y by establishing forecasts for aggregates of its subcomponents: consumption, investment, net exports, and government spending. Once these control totals have been established, adjustments to sectoral elements may be applied either through a proportional adjustment to a previous sectoral value or through some other procedure incorporating detailed current data on the movements of a few critical sectors.

Macromodeling of the movements of interrelated GNP components is a complicated challenge and may involve the regression estimation of a vast simultaneous equation system. Further complicating these estimations is the behavior of error terms in the different equations. If error terms are normally distributed, a least square approach of some kind may be useful. If, however, the error terms are distributed with weights in the tails of their distribution, other estimators may be more efficient. This is likely to be an important consideration in the estimation of parameters of a control equation for investment.

2. ROBUST REGRESSION

2.1 Modeling Assumptions, Outlier Observations, and the Estimation Problem

Assumptions on which statistical models are built may often be based on very limited knowledge. The extent to which the accuracy of a model may be affected by given deviations from assumptions is an important general question. "Robustness signifies insensitivity to small deviations from assumptions" (Huber, 1981, p. 1).

The field of robust statistics emerged as a recognized trend of formal study in the early 1960s. The concerns of this field are generally focused on assumptions regarding the shape of distribution functions of random variables in statistical models. A number of procedures have been devised to generate "robust" parameter estimates with satisfactory statistical properties.

The problem that outliers bring to regression analysis may be considered formally in the following way (Huber, 1981, p. 155-160). The classical linear regression specification involving p unknown parameters $\theta_1, \ldots, \theta_p$, independent variables x, a dependent variable y, and stochastic term u may be written as

$$y = \sum_{j=1}^{p} x_j \theta_j + u$$

The error term is assumed to be distributed with a mean of zero and constant finite standard deviation. Its realizations are assumed to be uncorrelated.

The unknown parameters are to be estimated from a set of n observations of y and x where the ith observation may be expressed as

$$y_i = \sum_{j=1}^{p} x_i \theta_j + r_i$$

where r_i is a residual in the ith observation.

Typically, the $_j$ are estimated by minimizing the sum of squared errors:

$$\sum_i y_i - \left(\sum_j x_{ij}\theta_j\right)^2 = \text{min!}$$

which is solved by differentiating with respect to the choice of θ and equating to zero:

$$\sum_i y_i - \left(\sum x_{ik}\theta_k\right)x_{ij} = 0$$

If X is a matrix of observations i on independent variables x_j and θ is a vector of parameter estimates, these estimates are established (given full rank of X) by the matrix relation

$$\theta = (X'X)^{-1}X'Y$$

where the prime denotes transposition. These estimates are unbiased and among the class of unbiased estimators are of minimum variance. They also correspond to the estimators derived by maximum likelihood methods if the residuals are assumed to be normally distributed.

The relation between the vector of observed values of y and the vector of fitted values $\hat{y}$ derived by using the parameter estimates is given by the relation

$$\hat{y} = X(X'X)^{-1}X'y = Hy$$

The matrix H is symmetric, idempotent, of dimension n × n, with p eigenvalues equal to 1 and n - p eigenvalues equal to 0. The diagonal elements h_{ii}, denoted here by h_i, are bounded by the relation

$$0 \leqslant h_i \leqslant 1$$

Note now that the ith residual can be written as

$$r_i = y_i - y_i = (1 - h_i)y_i - \sum h_{ik}y_k, \quad k \neq i$$

Consider the case in which there is a large realization u_i for the observation y_i. If h_i is close to 1, such a realization will not necessarily show itself in the residual r_i. Rather, it may appear in the calculation of some r_k for which h_{ki} happens to be large.

Points with large h_i values are known as leverage points and have an overriding influence on the determination of the fitted regression line in general and their own fit in particular.

As mentioned above, least squares estimators have the virtue of being unbiased and of having minimum variance within the class of unbiased estimators. They are also associated with the maximum likelihood estimators under the assumption that the disturbance terms are normally distributed.

Suppose, however, that the error terms are not normally distributed. Suppose, moreover, that the true disturbance distribution function is characterized by "greater weight in the tails" than the normal distribution. Samples, then, would be more likely to possess leverage points which, by their nature, are not obvious from a cursory look at fitted values. Although least squares estimators would remain minimum variance among the class of unbiased estimators, these might be less efficient than certain biased estimators which place less importance on outliers than squared error procedures.

2.2 Maximum Likelihood Type Estimates (M-Estimates)

The difference between an observed value of a dependent variable and the plotted value given by the vector multiplication of parameters and independent variables establishes a residual. Given as assumption regarding the distribution f( ) of such residuals and their independence, a likelihood function may be established:

$$L = \prod_{i=1}^{n} f(r_i) = \prod_{i=1}^{n} f\left(y_i - \sum x_i \theta_i\right)$$

for n observations. The method of maximum likelihood involves the choice of the $_i$ such that the likelihood of observing the sample which in fact was observed will be maximized. Typically, the procedure involves working with the log of the likelihood function, since ln L is often more tractable and does not alter the maximization problem.

Given the assumption of normally distributed errors, the maximum likelihood problem ends in the minimization of the sum of squared residuals—the least squares method.

Huber (1964) proposed a generalization of the procedure. Denote a parameter estimate of θ by $T_n(x_1, x_2, \ldots, x_n)$. An estimate T_n defined by the minimum problem

$$\sum \rho(x_i; T_n) = \min!$$

or implicitly by

$$\sum (x_{ij}; T_n) = 0$$

where ρ is an arbitrary function and $\psi(x;\theta) = (d/d\theta)\rho(x;\theta)$ is called an M-estimate or maximum likelihood-type estimate. If p(x;) takes the form of -log f(x;), the M-estimate is rendered as the ordinary maximum likelihood estimate.

The estimation of a location parameter is performed by working with some $\rho(x_i - T_n)$. In the case of regression, Huber (1981, p. 162) suggests that instead of minimizing a sum of squares, one minimizes "a sum of less rapidly increasing functions of residuals":

$$\sum_{i=1}^{n} \rho \; y_i - \sum x_{ij}\theta_j \quad = \min!$$

or, after taking derivatives,

$$\sum_{i=1}^{n} \psi\left(y_i - \sum x_{ij}\theta_j\right) x_{jk} = 0$$

3. ROBUST REGRESSION AND FINAL DEMAND CONTROLS

The final demand accounts of U.S. government input-output models may be segmented into separate consumption, investment, net export, and government spending accounts.

Forecasts of aggregate levels for these segments may be generated from a simultaneous equation system or from a set of independent equations. The most volatile component is likely to be that of investment and it is on this segment that attention now is focused.

Consider an autoregressive forecasting scheme

$$I_t = b_0 + b_1 I_{t-1} + b_2 R_t + u_t$$

where I represents aggregate gross domestic fixed investment, R represents an index of real interest rates, and u_t represents a disturbance term. Because of the historical volatility of the investment series, one cannot be sure that the u_t term will be normally distributed; indeed, it is fair to expect the disturbance term to be distributed with greater weight in its tails.

For this reason, robust estimation methods seem appropriate. Using a modest convex function ρ, an M-estimator for the coefficients b may be established and calculated through the use of iterative schemes presented by Huber (1981, pp. 179-192).

In this way, short-term forecasts for investment aggregate controls may be generated. Elements of the vector of the investment component of final demand may be adjusted proportionally to the forecast change in aggregate investment, or further information on the movement of key elements may be sought. Roughly half of all final demand investment is organized within the 85-industry Commerce Department input-output model into a single sector 11, new construction. This has been a stable relationship over the years during which the Commerce Department has prepared tables. Additional information on new construction may be applied to improve sectoral forecasts.

Investment sectoral forecasts, when added to sectoral forecasts for consumption, government spending, and net exports, would generate a final demand forecast which might usefully be applied in input-output studies.

REFERENCES

Huber, P. J. (1964). Robust estimation of a location parameter. Ann Math. Stat. 35: 73-101.

Huber, P. J. (1981). Robust Statistics. Wiley, New York.

Interindustry Economics Division, Bureau of Economic Analysis (1984). The input-output structure of the U.S. economy. Survey of Current Business, May, pp. 42-84. Department of Commerce, Washington, D.C.

Leontief, W. (1941). The Structure of American Economy 1919-1929. Harvard University Press, Cambridge, Mass.

Leontief, W., et al. (1953). Studies in the Structure of the American Economy. Oxford University Press, London.

Leontief, W. (1966). Input-Output Economics. Oxford University Press, London.

10

Using an Empirical Transformation Technique to Detect Outliers for Improved Accuracy in Forecasting Models

MICHAEL D. GUERTS and H. DENNIS TOLLEY Brigham Young University, Provo, Utah

1. INTRODUCTION

The purpose of this chapter is to present a method of identifying outliers in time series data. An outlier is an observation that is aberrant or unusually different from the rest of the observations. Often outliers arise in real data from known causes such as a change in pricing policy, a business promotion, or a labor strike. When the researcher can easily identify the reason for an outlier, the outlier itself can be identified. Once it is identified, steps to remove its influence on the time series model used for forecasting can be undertaken. This process of removing the influence of outliers makes the models robust to the effects of outliers.

The effect of outliers on modeling and prediction in the case of independent and identically distributed random variables in well established (see, e.g., Huber, 1981).

Of particular interest in time series modeling is the effect of an outlier on measures of autocorrelation. Devlin et al. (1975) showed that for some measures of correlation, estimations from an independent series of correlated pairs of random variables were greatly influenced by an outlier pair. Particularly Pearson's correlation, related to least squares-type estimation, was greatly biased by only a few outlier pairs.

Denby and Martin (1979) showed that similar biases occurred in simple first-order autoregressive processes. Others have shown effects of outliers on spectrum estimates (see Kleiner et al., 1979; Brillinger, 1973). Other impacts on autoregressive moving-average parameters have been studied by Denby and Martin (1979), Martin and Yohai (1984), and Lattin (1982).

The influence of an outlier is apparent in exponential smoothing forecasting methods. Assuming that the coefficient (α) used in the exponential smoothing is not estimated using an outlier observation, the effects of an outlier in the data will be averaged into succeeding forecasts made by exponential smoothing. Thus, the influence of the outlier will be successively downweighted for several forecasts until ostensibly removed by the exponential weight.

Unfortunately, most of the methods of identifying and removing outliers require (1) identification of the ARMA model order and (2) a possibly iterative method of estimating the parameters. We note that the method proposed by Lattin (1982) does not require identification of the model and parameter estimates but bases outlier identification on measures of autocorrelation. Therefore, with the exception of that work, methods for identifying and removing outliers may be too involved computationally to be of practical use in forecasting a time series.

In this chapter we propose a method of identifying outliers which is simple to implement and does not require complete identification of the model parameters. The method resembles, graphically, the quantile-quantile plots used in examining tail behavior from observations of independently identically distributed random variables. The procedure is easily implemented as a preprocessing step to a full time series analysis and can be used in conjunction with standard time series packages. It can also be used in conjunction with more ad hoc forecasting methods such as exponential smoothing as well as ARIMA methods. We illustrate the method using simulated time series with known dynamics. The method is then used to check for outliers in the Hawaii tourist data.

2. PRELIMINARIES

We will consider a time series which is a discrete-time ergodic normal process, denoted by X(n), n = ..., -1, 0, 1, ... This means that for any integral value of n, the probability distribution of X(n) is normal with some fixed mean μ and some fixed variance σ^2. Both μ and σ^2 are independent of n. Writing in probability terms, this means that for any real x

$$\text{Prob}\{X(n) \leq x\} = \Phi\left(\frac{x - \mu}{\sigma}\right) \tag{1}$$

Usually X(n) will have some familiar dynamic form such as an ARMA process. Note that equation (1) applies to a wider range of time series X(n) than simply ARMA models. Hence, although X(n) may be assumed to be an ARMA process, equation (1) is certainly not restricted to ARMA processes. Note, however, that X(n) must be ergodic.

To apply equation (1) to the identification of outliers, we need a definition of outliers for time series data. There are several ways to define and generate time series outliers. In the case of independent observations, the common scenario is that "most" of the observations are sampled from one distribution and the remainder from another. Mathematically, the random variable Y_n would be said to have distribution function F(y), where

$$F(y) = \alpha G_1(y) + (1 - \alpha)G_2(y), \quad 1/2 < \alpha \leqslant 1 \tag{2}$$

In equation (2) α represents the proportion of the population that is sampled from the population with distribution function $G_1(y)$. $G_2(y)$ is the distribution function for the outliers. Following similar logic, time series outliers can be defined by assuming that X(n) is contaminated by chance for certain values of n. Operationally, one observes the random time series Y(n), where Y(n) = X(n) + V_n, where V_n are independently distributed random variables which take on a value of zero with probability α and follow distribution $G_2(y)$ with probability $1 - \alpha$. Under this definition, any dynamic structure, such as an ARMA process, is not influenced by the outlier other than in contaminating the actual observations. Such outliers are often referred to as "additive outliers" (see Denby and Martin, 1979). Common examples of additive outliers are electrostatic shock for an electronic monitoring system, temporary change in airline fares to promote tourist trade, crop failure for agricultural futures, and so forth. Alternatively, outliers are defined as "innovative outliers," where each outlier enters into the dynamic framework generating the time series. Thus the effect of the outlier is promulgated through the series from that time point. Yet a third type of outlier is the type in which a temporary shift in the dynamic structure causes the data momentarily to be generated by a different process. In this chapter we consider only additive outliers.

Again, referring back to the independent observation case there are many methods for detecting outliers. One of the simplest to implement when the form of G_1 is known is the quantile-quantile plot. A quantile of a random variable is like a percentile. It is a point below which some fixed proportion, say q, of the population is expected to fall. For example, if q = .1, the quantile is the point below which 10% of the population lies. The q = .5 quantile corresponds to the median.

If the set of n observations is drawn from a known distribution, the quantiles $q_k = k/n$ can be determined in two ways. First, the known distributions will yield the true quantiles on theoretical grounds. Second, if $y_{(k)}$ represents the kth smallest of the n observations, then $y_{(k)}$ is the q_kth quantile of the sample, approximately the q_kth quantile of the population. If the assumed known distribution is, in fact, the true distribution, then the estimate $y_{(k)}$ will correspond to the true quantile except for random fluctuations. A plot of the $y_{(k)}$ values against the q_k quantiles calculated using the true distribution will approximately follow a 45° line as k = 1, ..., n. Large deviations of this plot of points from the 45° line indicate that the assumed distribution is incorrect. Thus, if one assumes the data follows $G_1(y)$, in equation (2), and $\alpha \neq 1$, then the outliers will show up as causing the plot to deviate from the 45° line. An extensive examination of quantile-quantile plots for distribution identification is given by Chambers et al. (1983).

3. IDENTIFYING ADDITIVE OUTLIERS IN TIME SERIES

The quantile-quantile plot can be used in the time series setting because of equation (1). To identify the outliers, order the observations X(1), ..., X(n) from the smallest to the largest, ignoring their temporal order. These observations are then plotted against the quantiles of a normal random variable with the same mean and variance. Outliers in the time series will cause the plot to deviate from the 45° line, particularly at the "ends" of the data. If the data have no obvious outliers but are simply nonnormal, the plot will show irregular behavior. This method of identifying outliers is called the empirical transformation methodology because of the use of the quantile-quantile plots.

As an example, consider a first-order autoregressive time series with parameter 0.3. Then the model for X(n) is

$$X(n) = 0.3X(n-1) + \xi_n \tag{3}$$

where ξ_n are independent and identically distributed normal random variables with mean zero and variance unity. Figure 1 gives a series of 100 simulated values using equation (3). [The starting point X(0) was simulated using equation (3) with X(-100) = 0.]

The quantile-quantile plot was calculated by first determining the sample mean and standard deviation of the data. In this case $\bar{X} = -0.40$ and S = 1.41. The quantiles k/100, were calculated for the standard normal distribution (mean = 0, variance = 1). These quantiles were then

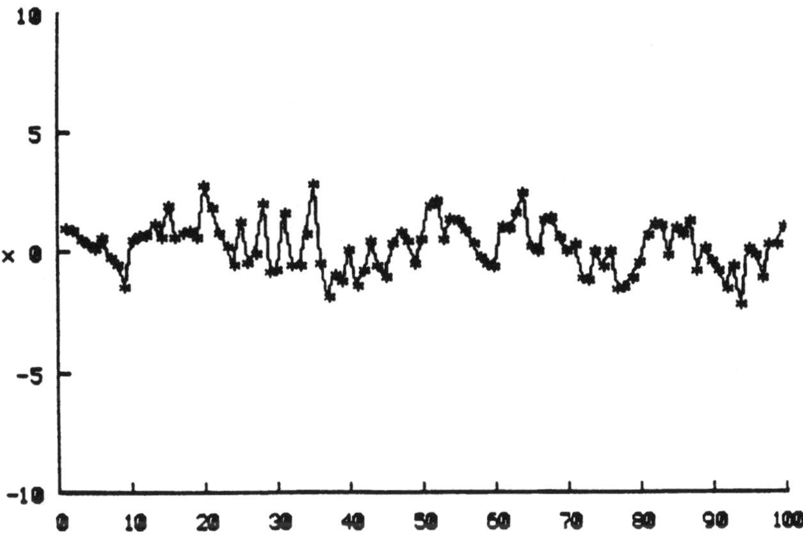

FIGURE 1 Plot of 100 observations simulated with a first-order autoregressive model with parameter $\phi = 0.3$. Noise is distributed as a normal random variable with mean = 0 and variance = 1.

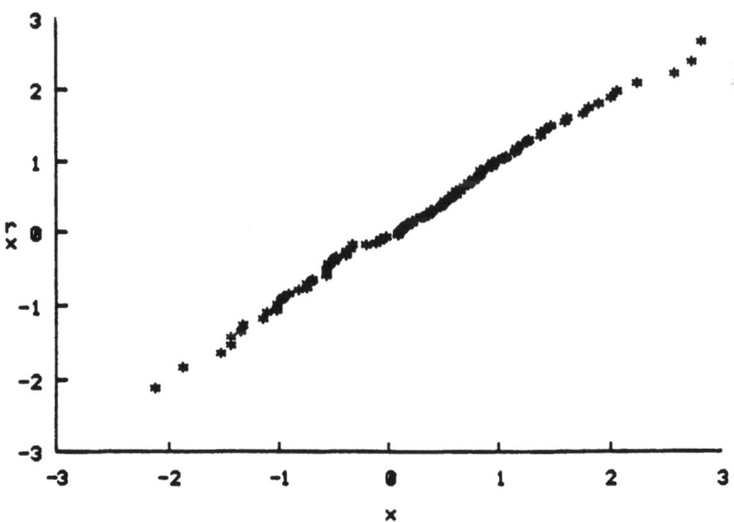

FIGURE 2 Quantile-quantile plot of the data from Figure 1.

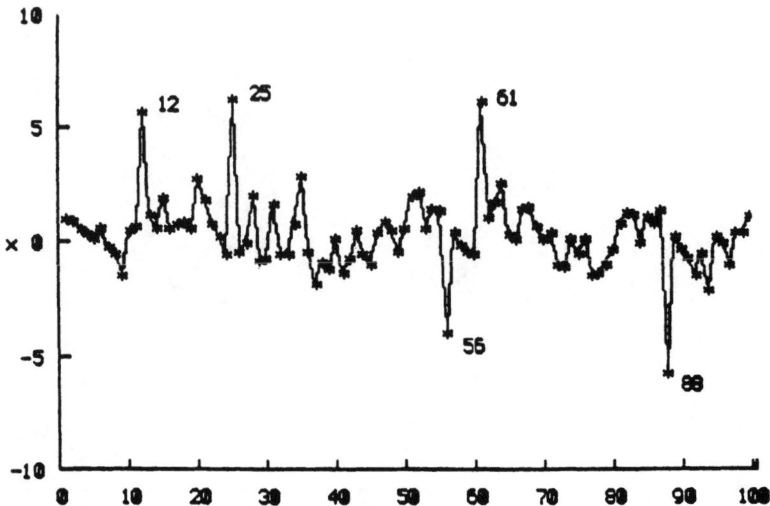

FIGURE 3 Plot of data from Figure 1 with five outlier observations added.

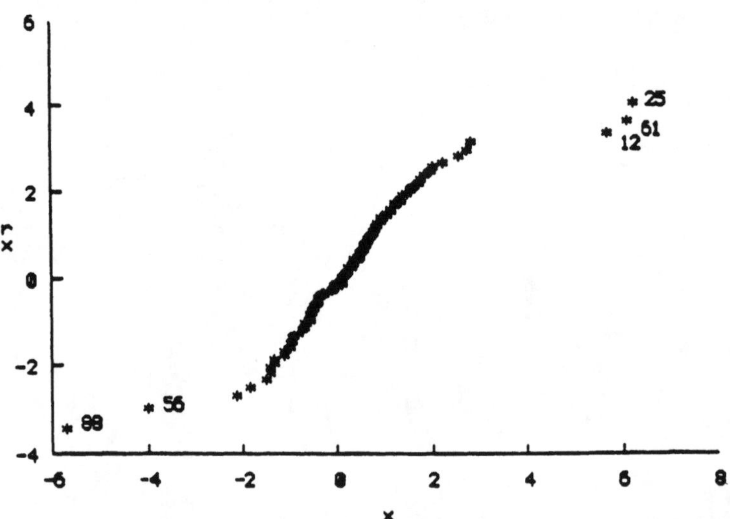

FIGURE 4 Quantile-quantile plot of data from Figure 3.

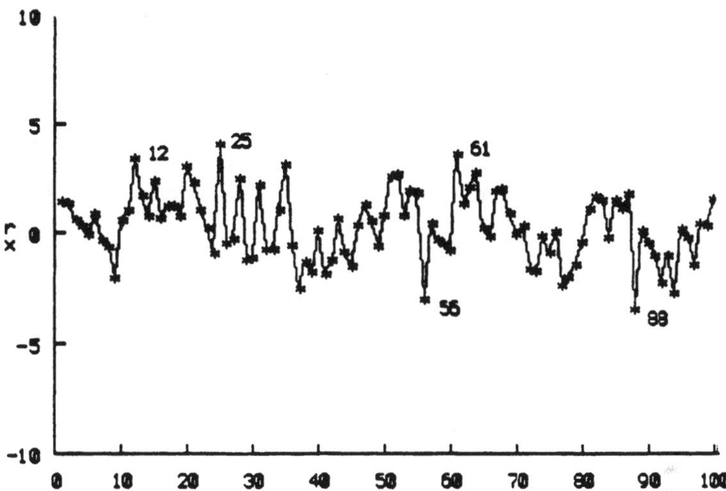

FIGURE 5 Time series plot of data in Figure 3 after all observed values are replaced by corresponding quantile values.

transformed to the same scale as X(n) by multiplying each quantile by S and then adding $\overline{X}$. The quantile-quantile plot is given in Figure 2 for these data. Note that the plotted points seem to fall approximately on a 45° line.

A time series process with outliers is generated from the data in Figure 1 by adding a constant to five points, creating five outliers. To X(12), X(25), and X(61) we added 5. To X(56) and X(88) we added -5. The resulting time series is given in Figure 3. Note that each outlier point has been identified with the corresponding value of n. Following the same steps as above, a quantile-quantile plot is generated for the data in Figure 3. This plot is given in Figure 4. The five generated outliers' points are clearly off the 45° line on which most of the data lie. The difference in the x-axis and y-axis scales make the 45° line actually appear steeper than a 45° line.)

Note that in forming the plots given in Figures 2 and 4, no attempt was made to remove the autocorrelation. As the autoregressive parameters become more important (farther from zero), stronger cyclic patterns emerge in the data. In our experience, the stronger the cyclic pattern, the longer the series must be to identify true outliers. If the series is short, the method may identify extreme observations resulting from a true harmonic as outliers. One way around this problem is to remove any strong high-frequency cyclic patterns in a preprocessing step.

We mention in passing that one may downweight the influence of any outliers by replacing the observed value of each X(n) with its quantile. This will keep all X(n) values in their same relative ordering with respect to neighboring observations but at the same time make the distribution of X(n) appear more normal. Such a transformation of the data shown in Figure 3 is given in Figure 5.

4. APPLICATION TO HAWAII TOURIST DATA

To illustrate the use of the methodology presented in Section 3, we will examine the Hawaii tourist data for outliers. This is done with actual data. The time series used represents overnight visitors to Hawaii. The tourist-to-Hawaii time series was used by Geurts and Ibrahim (1975) in a forecasting effort. Bunn (1979) also examined this time series. A reforecast examining the outliers in the same data was done by Geurts (1982). Geurts (1982) identified outliers in the data by reviewing historical events that would be expected to generate outliers. The identified outliers are given in Table 1. The Geurts (1982) study replaced outlier data with forecast values for the outliers and found that future forecasts were greatly improved in accuracy. The yearly sales are given in Figure 6 and the monthly sales are given in Figure 7.

A plot of the Hawaii tourist data revealed that the data violate the assumptions in three ways.

1. There is a seasonality component.
2. The variability increases over time.
3. The data have a trend in the mean over time.

In order to use the outlier identification procedure, we removed these problems as follows:

1. Seasonal effect was removed by estimating monthly seasonality factors. These were estimated using the first 60 months of data. The process of generating seasonal factors was to detrend the first 60 data points and then average the total and average each month. The ratio of the month's average divided by the average of all months created a seasonal factor for a month. Subsequent monthly observations were "seasonally adjusted" by dividing by these factors.
2. Changing variability effect was reduced by looking at the data in overlapping windows of 48-months length. Each window

TABLE 1 Atypical Periods: 1969-1978

Period	Event
1. March 1970	747 flights start.
2. September 1970 October 1970 November 1970	Island-wide hotel strike.
3. December 1970	Pollution on Waikiki Beach. National coverage.
4. January 1971	Bad mainland weather. Flights canceled.
5. December 1971 January 1972 February 1972	Nonaffinity group fare in effect. West coast fares reduced 35% ($70). Fare later eliminated (March 1972).
6. January 1973 February 1973	U.S. changes to Monday holiday for Washington's birthday. Hotels overbook. State institutes operation ALOHA. Tourists placed in private homes, sent home, or slept on cots in hotel lobbies.
7. August 1973	American Legion convention.
8. June 1975	Unknown.
9. November 1975 December 1975	American Medical Association convention.
10. June 1976	Lions convention.

overlapped the last 36 months of the previous window. Outliers identified in two or more windows were considered strong outliers.

3. For each window, the trend was removed by fitting a simple linear regression to the 48 data points. The data actually used in the quantile-quantile plot are the residuals from these fits.

Since the outliers have already been identified (see Table 1) from historical data, it is interesting to see if the techniques explained earlier in this chapter will identify the same outliers. The procedure was to take windows of length 48 months and make a quantile-quantile plot of the Hawaii data for this period. We started with 1971 data. Points that did

Date	Actual Sales
1948	36397.0
1949	34386.0
1950	46593.0
1951	51565.0
1952	60539.0
1953	80346.0
1954	91289.0
1955	109798.0
1956	133815.0
1957	168829.0
1958	171588.0
1959	243216.0
1960	296517.0
1961	319807.0
1962	362145.0
1963	429140.0
1964	563925.0
1965	686928.0
1966	835456.0
1967	1124818.0
1968	1314571.0
1969	1527012.0
1970	1746970.0
1971	1818944.0
1972	2244377.0
1973	2630952.0
1974	2786489.0
1975	2829105.0
1976	3220151.0
1977	3433667.0
1978	3670309.0
1979	3960531.0
1980	3934504.0
1981	3934623.0
1982	4242925.0
1983	4367880.0
1984	4855580.0

FIGURE 6 Yearly visitors to Hawaii.

not lie on the 45° line were considered outliers. The more frequently a period appeared as an outlier in a window, the higher the probability that it was indeed an outlier each data point would occur a maximum of four times in this study. Table 2 is a listing of the outliers in the Hawaii

DATE	ACTUAL SALES
1975	
JAN	233987.0
FEB	238621.0
MAR	259725.0
APR	195839.0
MAY	201601.0
JUN	254443.0
JUL	268739.0
AUG	285964.0
SEP	220690.0
OCT	222225.0
NOV	219538.0
DEC	227733.0
1976	
JAN	253326.0
FEB	256201.0
MAR	265860.0
APR	259490.0
MAY	245402.0
JUN	286129.0
JUL	314331.0
AUG	322509.0
SEP	242065.0
OCT	254521.0
NOV	243500.0
DEC	276817.0
1977	
JAN	298028.0
FEB	266542.0
MAR	310494.0
APR	281069.0
MAY	253017.0
JUN	300092.0
JUL	312457.0
AUG	331399.0
SEP	251448.0
OCT	272929.0
NOV	252777.0
DEC	303415.0
1978	
JAN	319267.0
FEB	280517.0
MAR	325412.0
APR	284756.0
MAY	266680.0
JUN	314227.0
JUL	337085.0
AUG	331647.0
SEP	274725.0
OCT	294835.0
NOV	305139.0
DEC	336019.0
1979	
JAN	370870.0
FEB	328598.0
MAR	361362.0
APR	291735.0
MAY	260867.0
JUN	332622.0
JUL	363080.0
AUG	394608.0
SEP	298766.0
OCT	311924.0
NOV	301256.0
DEC	344843.0
1980	
JAN	336349.0
FEB	305512.0
MAR	362969.0
APR	308580.0
MAY	297709.0
JUN	336564.0
JUL	392511.0
AUG	378865.0
SEP	275906.0
OCT	286470.0
NOV	299969.0
DEC	353100.0
1981	
JAN	307518.0
FEB	320294.0
MAR	331229.0
APR	313307.0
MAY	318320.0
JUN	352847.0
JUL	357872.0
AUG	394199.0
SEP	290259.0
OCT	308013.0
NOV	296854.0
DEC	343911.0
1982	
JAN	336520.0
FEB	337970.0
MAR	379555.0
APR	343890.0
MAY	327395.0
JUN	377390.0
JUL	397870.0
AUG	398405.0
SEP	314455.0
OCT	344510.0
NOV	342195.0
DEC	342770.0
1983	
JAN	325385.0
FEB	348290.0
MAR	396270.0
APR	318900.0
MAY	341215.0
JUN	411395.0
JUL	410900.0
AUG	418135.0
SEP	317370.0
OCT	351105.0
NOV	329780.0
DEC	399360.0
1984	
JAN	378980.0
FEB	396680.0
MAR	442800.0
APR	383320.0
MAY	388860.0
JUN	434630.0
JUL	431450.0
AUG	446220.0
SEP	345270.0
OCT	390550.0
NOV	383810.0
DEC	433010.0

FIGURE 7 Monthly visitors to Hawaii.

TABLE 2 Table of Outliers

Month	Number of occurrences	Windows of occurrence
Jun 72	1	72-75
Feb 73	1	71-74
Mar 74	2	72-75, 74-77
Jun 75	3	72-75, 73-76, 74-77
Jan 76	1	73-76
Jan 77	1	74-77
Mar 77	1	74-77
Aug 78	4	75-78, 76-79, 77-80, 78-81
Jan 79	4	76-79, 77-80, 78-81, 79-82
Jun 79	1	76-79
Jan 80	1	80-83
Jun 80	3	77-80, 79-82, 80-83
Aug 84	1	81-84
Nov 84	1	81-84
Jan 85	1	82-85
Jun 85	1	82-85

data as determined by the empirical transformation method. Figures 8 through 13 show the plots to determine the outliers. Ten outliers were identified by historical analysis. The empirical transformation method identified two of the same months as outliers: February 1973 (one window) and June 1975 (three windows). For the 1969-1978 period, the empirical transformation identified three additional months as being outliers that were not identified by the historical analysis. These were June 1972, March 1974, and January 1976. Perhaps the historical analysis overestimated the number of outliers. In combination, both techniques identified 14 of the 84 months as outliers. Only two months were identified by both techniques.

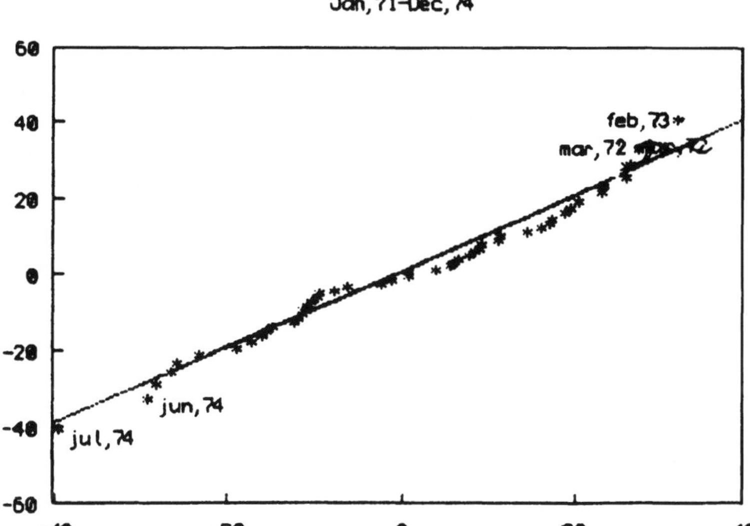

FIGURE 8 Monthly visitors to Hawaii.

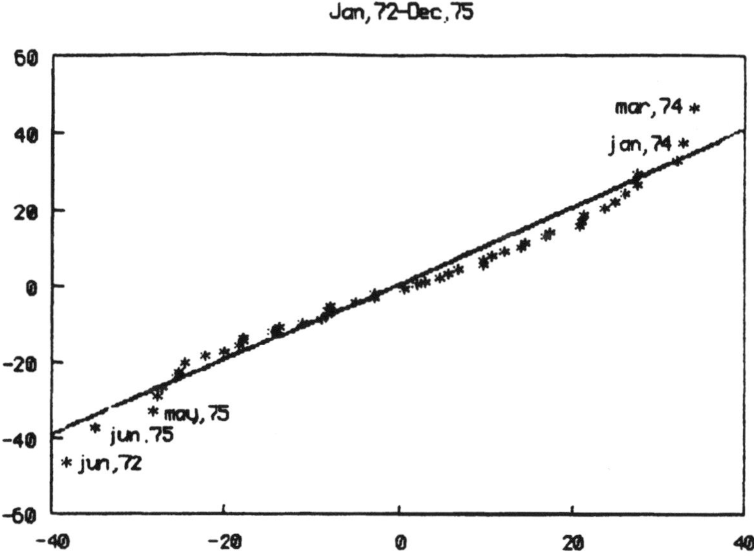

FIGURE 9 Monthly visitors to Hawaii.

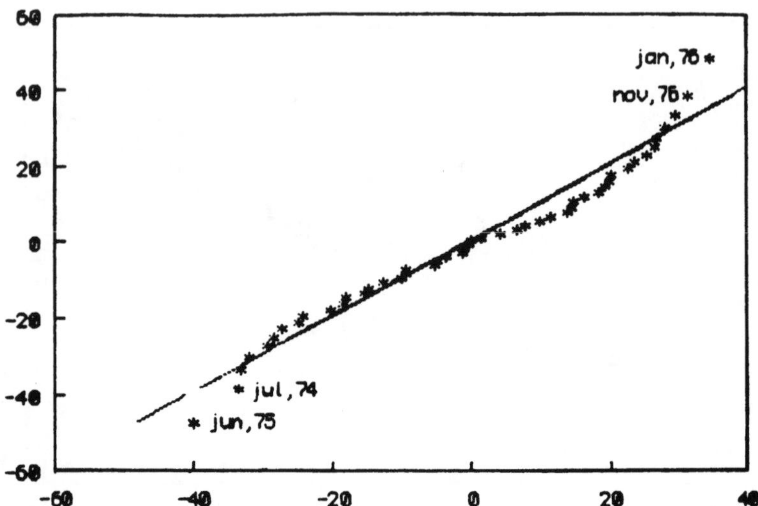

FIGURE 10 Monthly visitors to Hawaii.

Jan,74-Dec,77

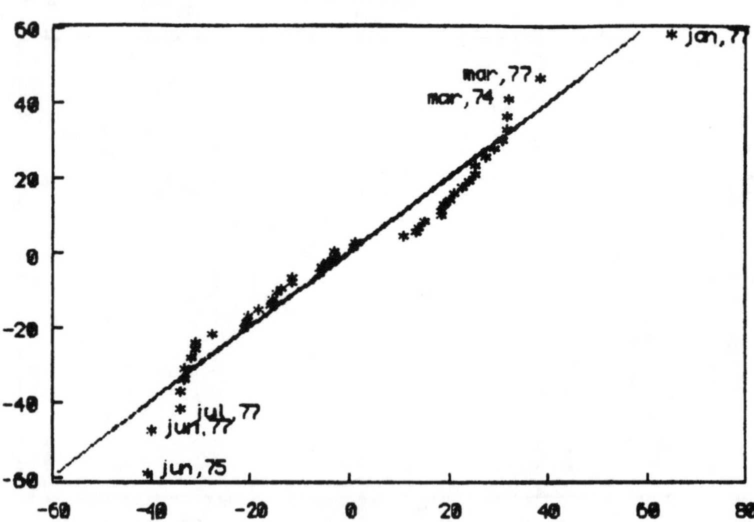

FIGURE 11 Monthly visitors to Hawaii.

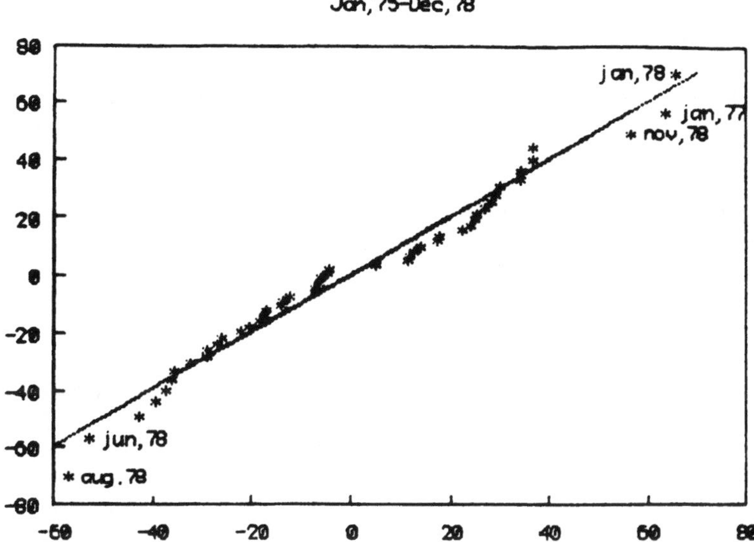

FIGURE 11 Monthly visitors to Hawaii.

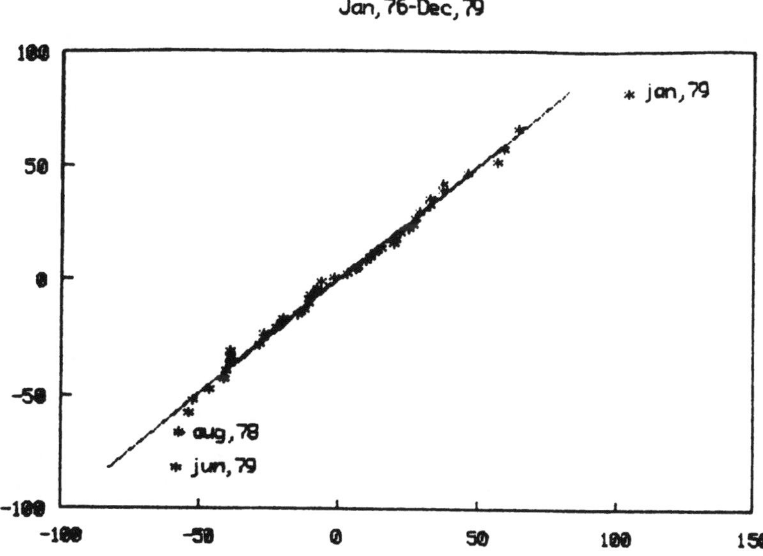

FIGURE 13 Monthly visitors to Hawaii.

The historical evalution leaves much to be desired. For example, from Table 1, the start of 747 flights was thought to cause an outlier. The empirical transform did not identify such an outlier. In effect, the 747 flight may have had no effect or perhaps only a modest effect on travelers. Also, a hotel strike may inconvenience travelers but may not keep them from coming. Conventions may not cause increased travel but may only replace ordinary tourists with conventioneers. An empirically based analysis of outliers clearly examines the "outlier" from a different perspective than a historical analysis.

5. DISCUSSION

In this chapter we have noted that a very large improvement in forecasting accuracy can be gained by identifying and then adjusting the outliers. The empirical transform outlier identification technique provides an easily implemented preprocessing step in identifying such outliers. Clearly, one never knows whether a particular observation is a true outlier. However, through the graphical methods illustrated in this chapter, one can see which observations appear to deviate from expectation. These graphical techniques can be used on an ergodic process without removing autocorrelations first. This is an extremely fortunate result, since estimating autocorrelation is problematic when outliers are present. We note also that the method of identifying outliers suggests a method of reducing their effect. Explicitly, if each observed value is replaced by its expected quantile, the order relationship in the data is maintained with outliers "eliminated."

REFERENCES

Brillinger, D. R. (1973). A power spectral estimate which is insensitive to transients. Technometrics 15: 559-562.

Chambers, J. M., Cleveland, W. S., Kleiner, B., and Tukey, P. A. (1983). Graphical Methods for Data Analysis. Duxbury Press, Boston.

Denby, L., and Martin, R. D. (1979). Robust estimation of the first order autoregressive parameter. J. Am. Stat. Assoc. 74: 140-614.

Devlin, S. J., Gnanadesikan, R., and Kettenring, J. R. (1975). Robust estimation and outlier detection with correlation coefficients. Biometrika 62: 531-545.

Franke, J., Hardle, W., and Martin, D., eds. (1984). Robust and Nonlinear Time Series Analysis. Springer-Verlag, New York.

Geurts, M. D. (1982). Forecasting the Hawaiian tourist market. J. Travel Res. (Summer).

Geurts, M. D., and Ibrahim, I. B. (1975). Comparing the Box-Jenkins approach with exponentially smoothed forecasting model application to Hawaii tourists. J. Market. Res. (May).

Huber, P. J. (1981). Robust Statistics. Wiley, New York.

Kleiner, B., Martin, R. D., and Thomson, D. J. (1979). Robust estimation of power spectra. J. R. Stat. Soc. B. 41: 313-351.

Lattin, J. M. (1982). Identifying influential observations in time series data. Proceedings, 14th Annual Symposium on the Interface.

Martin, R. D., and Yohai, V. J. (1984). Gross error sensitivities of GM and RA-estimates. Robust and Nonlinear Time Series, J. Franke, W. Hardle, and D. Martin, eds. Springer-Verlag, New York.

11

A New Look at Mergers in the United States, 1895-1973

JOHN B. GUERARD, Jr. Drexel Burnham Lambert, Chicago, Illinois

KENNETH D. LAWRENCE Rutgers University, Piscataway, New Jersey

1. INTRODUCTION AND LITERATURE REVIEW

Using quarterly data from 1895 to 1904, Nelson (1959) found a correlation coefficient of .613 between mergers and stock prices. The positive correlation between mergers and stock prices should exist because business executives are more willing to merge their businesses when the stock prices they receive are increasing. The acquiring firm's managers are more able to pay the higher prices for the acquired firm's shares as the price-earnings multiple of the combined entity rises. Financial theory was developed by Larson and Gonedes (1969) to explain the conglomerate merger movement of the 1960s in terms of the price-earnings multiples. Larson and Gonedes hypothesized that although the price-earnings multiples of the acquiring firms exceeded those of the acquired firms, the market should drive the price-earnings multiples of the combined entity to the weighted average of the constituents' precombination earnings multiples. The incremental value of the combined entity would be zero if the combined entity's price-earnings multiple equaled the premerger weighted average multiple. The lack of merger profits is evidence of the perfectly competitive acquisitions market (PCAM), according to which the price paid for the acquired forms is such that the acquiring firms will not profit (Mandelker, 1974). Empirical evidence supports the existence of larger price-earnings multiples for nonacquired forms than for acquired firms in the premerger period (Harris et al., 1982). However, the acquiring firms' prices (and price-earnings multiples) in the postmerger period do not reflect merger profits and are consistent with the PCAM (Guerard, 1982). Beckenstein (1979) studied large mergers that occurred during the 1948-1975 period and found a positive coefficient on the stock price index variable in the merger equation; however, Beckenstein dismissed the association of stock price and mergers because of the numerically insignificant value. A similar conclusion was reached by Melicher et al. (1983), who found that stock prices led mergers by only one quarter. Because merger negotiations generally precede consummations by two quarters, the multiple time series association did not seem reasonable in a business environment. Salter and Weinhold (1982) noted that merger activity tends to increase as the ratio of market value to replacement value decreases. One might not expect a positive relationship between stock prices and mergers, given this reexamination of Tobin's q.

Guerard (1985) applied the Ashley et al. (1980) bivariate modeling test to the original Nelson data from 1895 to 1954 and found a positive and statistically significant relationship between mergers and stock prices. Moreover, statistical causality was found in stock prices producing mergers. Thus, the empirical evidence on stock prices and mergers is mixed.

The Nelson study also found a correlation coefficient of .259 between mergers and industrial production for the same time period. Mergers should increase as economic activity increases. Guerard (1985, 1989) found no causality association between mergers and industrial production. This study tests the existence of any statistically significant correlation among mergers, stock prices, and industrial production, using annual data from 1895 to 1973.

2. THE DATA

It is well known that the first merger movement, beginning in 1879 and ending in 1905, which coincided with the development of the modern capital market and a rising stock market, was characterized by large horizontal mergers. The second merger movement, lasting from 1916 to 1929, created oligopolies through vertical and conglomerate mergers. The third merger movement began in 1940 and reached its peak, in terms of number of mergers, in 1969 and has been primarily conglomerate in nature. It is interesting that the three major merger movements, although differing in the nature of the mergers, have been associated with rising stock market values and levels of industrial production.

The reader is referred to Stigler (1950), Nelson (1959), Lintner (1971), and Niemi (1975) as standard references for summaries of the merger history of the United States.

Regression analysis is applied to the annual series of mergers (M), the Dow Jones industrial average (DJIA), and industrial production (IP). The series are (1) the original merger series found in the Nelson study reflecting firm disappearances during the 1895-1954 period, (2) the index of the Dow Jones industrial average which was estimated with the arithmetic mean of the 12 monthly indices for a given year, and (3) an annual industrial production series developed in the Nelson study. Data from the 1954-1973 period were collected from the U.S. Federal Trade Commission and various volumes of the Economic Report of the President. The merger, stock prices, and industrial production series employ the logarithmic transformation because the plot of the range versus the mean of each subset of 20 observations reveals a relatively random scatter about a straight line (Jenkins, 1979).

3. ORDINARY LEAST SQUARES REGRESSION

Ordinary least squares (OLS) regression analysis of the merger series during the 1895-1973 period, using stock prices and industrial production as the independent variables, is plagued by severe multicollinearity. The correlation coefficient between stock prices and industrial production is .948. See Table 1 for the correlation matrix. It is not surprising that the regression coefficient on the industrial production variable is negative, given the extreme multicollinearity (Mason et al., 1975). The OLS results, when corrected for autocorrelation using the Cochrane-Orcutt (CORC) procedure, show that only stock prices are positively and significantly associated with mergers. The OLS equation is summarized in Table 4.

TABLE 1 Correlation Matrix, 1895-1973

	M	SP	IP
M	1.000	.797	.727
SP	.797	1.000	.948
IP	.727	.948	1.000

4. AN INTRODUCTION TO LATENT ROOT REGRESSION ANALYSIS

It is well known that the presence of multicollinearity produces unstable regression coefficients with inflated standard errors (Gunst et al., 1976). The standard OLS regression model is

$$Y = 1B_0 + XB + E$$

where

Y is an n × 1 vector of observable variables
B_0 and B^1 are unknown parameters
1 is an n × 1 vector of ones
X is an n × k matrix of standardized known independent variables
E is an n × 1 vector of random variables

The least squares estimator B is well known:

$$\hat{B} = (X^1X)^{-1}X^1Y$$

The expected value of the OLS estimator is B, the true value, minimum-variance unbiased estimator. Multicollinearity, the condition of highly correlated independent variables, produces an almost (exact) linear dependence among the independent variables. The near singularity of the independent variables produces inflated standard errors of regression coefficients because the ill-conditioning distorts the $(X^1X)^{-1}$ calculation. Biased regression techniques, such as ridge regression and latent root regression (Gunst et al., 1976; Webster et al., 1974), were developed in an attempt to obtain stable regression coefficients.

Whereas ridge regression is concerned with the determination of the biasing parameter k, whether by coefficient ridge plotting or by estimation with an iterative procedure (Montgomery and Peck, 1982), latent root regression seeks to identify near singularities in the independent variables, determine the predictive value of the near singularities, and estimate modified regression coefficients adjusted for non-predictive near singularities.

The correlation matrix (A^1A) of dependent and independent variables has latent roots, λ_i, and latent vectors, α_{0i}, defined by

$$|A^1A - \lambda_i I| = 0$$

$$(A^1A - \lambda_i I)\alpha_i = 0$$

The OLS estimator of B can be written (Gunst et al., 1976) as

$$B = -n \sum_{n=0}^{k} a_i \alpha_i$$

and

$$a_i = \frac{\alpha_{0i} \lambda_i^{-1}}{\sum_{r=0}^{k} \alpha_{or}^2 / \lambda_r}$$

Multicollinearity is present when the latent roots and vectors are near zero. Exact linear dependence exists among the independent variables when some $\lambda_i = 0$; however, in practice, it is very rare for some $\lambda_i = 0$, but some λ_i may be very small, indicating near singularities. Webster et al. (1974) found that near singularities are present when the latent roots $\leq .30$ and latent root vectors $\leq .10$.

In a geometric interpretation, the latent root of a particular latent vector measures the spread of data points in the direction defined by the latent vector. The latent root represents the sum of squares and a small latent root indicates little variability in the orthogonal axis direction (Webster et al., 1974). A small latent vector means that the orthogonal axis is nearly orthogonal to the independent variable axis. Small latent roots and variables reveal nonpredictive near singularities.

The latent root regression (LRR) estimator should dominate the OLS estimator when the collinear variables produce coefficients in which a linear combination of vectors is orthogonal to the elements of the latent vector corresponding to the smallest latent root of the correlation matrix, V_1. That is, the subvector of regression coefficients, B, whose elements correspond to collinear variables, is a small multiple of the elements of the latent vectors of the smallest latent root. Latent root regression adds a biased term while removing the ill-conditioning. As $B^1 = cV^1$ and $|c|$ is small, the bias term is small and the mean square error of the latent root regression estimator is less than the mean square error of the ordinary least squares estimator. Thus, LRR analysis is preferred to OLS analysis as long as the parameter vector is not parallel to the latent vector corresponding to the smallest latent root of the correlation matrix.

5. A ROBUST REGRESSION WEIGHTING SCHEME

The ordinary least squares regression analysis produces no outliers (observations not within two standard deviations of the regression lines) in the annual regression; given a normal distribution of 79 observations, one would have expected at least three or four observations to lie outside the confidence intervals. Although there were no "actual" outliers, five observations were almost identical to the 95% upper confidence interval estimate for the observation. One could use the Beaton-Tukey (1974) biweight (robust) procedure for iteratively reweighting the regressions. Large residuals lead to very small observation weights. The biweight function is

$$w_i = [1 - (r/B)^2]^2 \quad \text{if } |r| \leq B$$
$$= 0 \quad \text{otherwise}$$

TABLE 2 Beaton-Tukey Weighting Scheme

Year	Weight[a]
1896	.932
1899	.880
1900	.960
1901	.959
1902	.961
1905	.975
1914	.975
1921	.964
1926	.979
1939	.970
1969	.949
1970	.976

[a] All other observations have a weight of 1.000 in the robust-weighted regression.

where

r = absolute value (residual/standard deviation of error)
B = a timing constant, 4.685

The optimal observation weighting scheme is shown in Table 2. It is interesting to note that only 1899 has an observation weight of less than .949, which is rather surprising given the merger activity of the 1960s. The robust-weighted OLS regression produces a positive and statistically significant coefficient on the stock price variable and an insignificantly negative coefficient on the industrial production variable. See Table 4 for complete robust-weighted OLS regression results.

6. LATENT ROOT REGRESSION RESULTS

Robust latent root regression analysis is a tool that might be useful in analyzing mergers, given the high degree of collinearity. The eigenvalues and eigenvectors for the 1895-1973 robust-weighted data are shown in Table 3.

The application of latent root regression techniques, using the vector deletion criteria of Gunst et al. (1976) and Webster et al. (1974) in which eigenvectors are deleted if the eigenvalues $\leqslant .30$ and eigenvectors are approximately $\leqslant .10$, leads to deletion of the third eigenvector of the robust-weighted data correlation matrix. The variance inflation factors are substantially reduced by the deletion of the third eigenvector. The latent root regression equation is summarized in Table 4. Both stock prices and industrial production are positively and significantly associated with mergers.

The application of biased estimator techniques produces positive regression coefficients on the stock price and industrial production variables in the annual merger regression equation for the

TABLE 3 Eigenvalues and Eigenvectors, 1895-1973

	1	2	3
Eigenvalues	2.65	.30	.04
Eigenvectors	.55	-.83	.14
	.60	.27	-.75
	.58	.50	.64

TABLE 4 Complete Regression Results

Equation	Dependent variable	(t) Constant	SP	IP	R^2
OLS	M	(.36) .173	(4.94) 1.427	(-1.27) -.400	.644
CORC	M	(.32) .095	(3.00) 1.078	(-.02) -.007	.281
Robust, OLS	M	(.30) .144	(4.94) 1.426	(-1.24) -.391	.644
Robust, LRR	M	.773	(7.27) .347	(12.60) .769	.576

1895-1973 period. Thus, it appears that one should not ignore the possibility of using robust-weighted biased estimation techniques in modeling time series merger models.

7. CONCLUSION AND SUMMARY

Robust-weighted biased estimation techniques are useful in analyzing mergers during the 1895-1973 period because of the high degree of collinearity between stock prices and industrial production. Unbiased regression techniques produce a negative coefficient on the industrial production variable. Latent root regression produces positive and significant coefficients on the stock price and industrial production variables. The coefficient on the stock price variable declines as one proceeds from OLS to latent root regression; however, the stock price coefficient does not appear to be numerically insignificant as Beckenstein found in his analysis of large mergers during the 1948-1975 period. Moreover, the positive association between the stock price and the merger variables casts doubt on the hypothesis that mergers occur because the firms are valued in the marketplace for less than the replacement cost of their tangible assets [the notion of Tobin's q (Salter and Weinhold, 1982)].

ACKNOWLEDGMENTS

The author appreciates the use of the latent root regression package developed by Professors S. Sharma and W. James. The robust-weighting scheme was estimated using SAS programs. Access to SAS was kindly provided by Air Products and Chemicals, Inc.

REFERENCES

Ashley, R., Granger, C. W. J., and Schmalensee, R. (1980). Advertising and aggregate consumption: An analysis of causality. Econometrica 48: 1149.

Beaton, A. E., and Tukey, J. W. (1974). The fitting of power series, meaning polynomials, illustrated on band-spectroscopic data. Technometrics 16: 147.

Beckenstein, A. R. (1979). Merger activity and merger theories: An empirical investigation. Antitrust Bull. 24: 105.

Butters, J. K., Lintner, J., and Cary, W. L. (1951). Corporate Mergers. Graduate School of Business, Harvard University, Boston.

Economic Report of the President. U.S. Government Printing Office, Washington, D.C.

Gort, M. (1969). An economic disturbance theory of mergers. Q. J. Econ. 93: 724.

Guerard, J. (1982). The role of employment and capital expenditures in the merger and acquisition process. Mergers and Acquisitions: Current Problems in Perspective, M. Keenan and L. White, eds. Lexington Book, Lexington, Mass.

Guerard, J. (1985). Mergers, stock prices, and industrial production: An empirical test of the Nelson hypothesis. Time Series Analysis: Theory and Practice 7, O. Anderson, ed. North-Holland, Amsterdam.

Guerard, J. (1989). Mergers, stock prices, and industrial production. Economics Letters.

Gunst, R. F., Webster, J. T., and Mason, R. L. (1976). A comparison of least squares and latent root regression estimators. Technometrics 18: 75.

Harris, R. S., Stewart, J. F., and Carleton, W. T. (1982). Financial characteristics of acquired firms. Mergers and Acquisitions:

Current Problems in Perspective, M. Keenan and L. White, eds. Lexington Books, Lexington, Mass.

Hoerl, A. E., and Kennard, R. W. (1975). Ridge regression: Iterative estimation of simulations. Commun. Stat. A5: 77.

Hoerl, A. E., and Kennard, R. W. (1970). Ridge regression: Biased estimation for nonorthogonal problems. Technometrics 12: 55, 66.

Hoerl, A. E., Kennard, R. W., and Baldwin, K. R. (1975). Ridge regression: Some simulations. Commun. Stat. 4: 105.

Larson, K. D., and Gonedes, N. J. (1969). Business combinations: An exchange ratio determination model. Accounting Rev. 44: 720.

Lintner, J. (1971). Expectations, mergers, and equilibrium in purely competitive markets. Am. Econ. Rev. 61: 101.

Lorie, J. H., and Hamilton, M. T. (1973). The Stock Market: Theories and Evidence. Richard D. Irwin, Homewood, Ill.

Mandelker, G. (1974). Risk and return: The case of merging firms. J. Financial Econ. 1: 303.

Mason, R. L., Gunst, R. F., and Webster, J. T. (1975). Regression analysis and problems of multicollinearity. Commun. Stat. 4: 277.

Melicher, R. W., Ledolter, J., and D'Antonio, L. J. (1983). A time series analysis of aggregate merger activity. Rev. Econ. Stat. 65: 423.

Montgomery, D. C., and Peck, E. A. (1982). Introduction to Linear Regression Analysis. Wiley, New York.

Nelson, R. (1959). Merger movements in American Industry, 1895-1956. Princeton University Press, Princeton, N. J.

Niemi, A. W., Jr. (1975). U.S. Economic History. Rand McNally College Publishing Company, Chicago.

Salter, M., and Weinhold, W. (1982). What lies ahead for merger activities in the 1980s. J. Business Strategy 2: 30.

Stigler, G. J. (1950). Monopoly and oligopoly by merger. Am. Econ. Rev. 40: 23.

U.S. Federal Trade Commission Staff Report. (1981). Statistical Report on Mergers and Acquisitions, 1979. Bureau of Economics, Washington, D.C.

Vinod, H. D., and Ullah, A. (1981). Recent Advances in Regression Methods. Marcel Dekker, New York.

Vinod, H. D. (1978). A survey of ridge regression and related techniques for improvements over ordinary least squares. Rev. Econ. Stat. 58: 121.

Webster, J. T., Gunst, R. F., and Mason, R. L. (1974). Latent root regression analysis. Technometrics 16: 513.

12

Alternative Methods of Dealing with Outliers in Forecasting Sales with Regression-Based Models

MICHAEL D. GEURTS and HEIKKI J. RINNE Brigham Young University, Provo, Utah

SHEILA M. LAWRENCE Rutgers University, North Brunswick, New Jersey

1. INTRODUCTION

Sales time series are frequently subjected to distortions caused by advertising, competition, promotions, accounting systems, and stockouts. These distortions are called outliers in statistical literature. In recent years, there has been an increasing number of articles dealing with the forecasting of sales using better time series models. In addition to building better models, a second way to improve forecasting accuracy is to improve the source data by a careful analysis of the process which generated the data.

Improving forecasting accuracy by analyzing data and improving the accuracy of the data is the topic of this chapter. The problem of bad data causing problems with statistical analysis has been discussed by Tukey (1977). Fisher (1966) discussed the problem of bad data for economic analysis. Krueger (1980) has examined the effects of disruptive events on forecasting seasonal time series. This chapter focuses on outliers and improving forecasting when outliers exist in the data. There are several ways of categorizing outliers. Outliers can be classified by their source:

1. The randomness of the underlying process
2. Acts of nature such as floods or fires
3. Human activities such as accounting techniques, promotions, or price cuts

Outliers can be classified as predictable, such as those due to promotions, and unpredictable, such as those due to randomness in the data.

An interesting problem is how to define outliers. For our present purposes, an outlier is defined as any value that is more than two standard deviations from the expected value. The focus of this chapter is the outliers that can be predicted, such as a promoted period.

An ARIMA model also can be written as a regression equation (see Makridakis et al., 1983). For example, the (1,1,1) ARIMA model can be written as

$$x_t = (1 + \phi_1)x_{t-1} - \phi_1 x_{t-2} + \mu + e_t - \phi_1 e_{t-1}$$

It can be seen from the regression form of the ARIMA and exponential smoothing models that an outlier can have an impact on the forecasting accuracy as forecasting progresses over time.

Distorted data affect forecasting accuracy in two ways. First, the forecaster may identify the wrong model or specify the wrong parameter values for the model as a result of using distorted data. Second, the upcoming forecast value for the period affected by a promotion will have a larger error than necessary unless the forecast is modified to reflect the atypical event that will produce the outlier.

2. DATA COLLECTION SOURCES OF OUTLIERS

2.1 Closing Books

Outliers can be caused by accounting methods. Some firms do not close their books for the month until all orders received during the month have been shipped. If an order is received on September 30 and is not filled until October 8, then September is not closed until October 8, and orders received October 1 and shipped October 3 are recorded as sales in September. The next year, September may be closed on October 2, six days earlier than in previous years. The result is that September may be 30 days one year, 40 days the next year, and 26 days the third year.

2.2 Crediting Returned Merchandise

Another Problem for forecasters caused by accounting methodology is the procedure the accounting department uses in handling returned merchandise. Returned products can be handled by accounting systems in two ways that result in a distortion of the data pattern that forecasters need to deal with. The first occurs when returned merchandise is credited to the month in which the return occurs, reducing sales for the month in which the merchandise was sold. This is a particularly difficult problem with seasonal merchandise. Christmas merchandise sold in December and returned in January may often cause January sales to be reduced substantially, and for some products a negative sales value may occur in January.

The second problem can occur if returned merchandise is credited by issuing a credit memo to the customer. If the credit memo carries a product number on it for the merchandise returned, and the credit memo is used to purchase a different product, some accounting systems will pick up the product number of the returned merchandise and record a sale of the product returned, rather than the product shipped. This overstates the quantity sold of the returned product and understates the sales of the product purchased with the credit memo.

2.3 The Number of Days in the Accounting Period

An assumption of time series forecasting is that the data are divided into equal periods. Often, accounting practices make the intervals unequal in length, and this can artificially distort the seasonal and trend patterns in the data.

A case in point is the accounting system in which the first and second months are 4-week months while the third month is a 5-week month. Accountants use this pattern when they want the books to close on the same day of the week. As a result, year after year, January will have 4 weeks, February 4 weeks, and March 5 weeks, and each month will end on a Friday. Unfortunately, the system requires that every third year, one 4-week month becomes a 5-week month. If, for example, July is a 4-week month for two years, every third year it becomes a 5-week month. (The reason for the extra week is that 356 days divided by 7 is 52.143. There are over 52 weeks in a year.)

The 4, 4, 5-week accounting procedure is obviously not an equal-interval time series. Its use will induce outliers in the sales data, buying until the promoted period. They reduced safety stock and made it up during the promoted periods. Then, as the promotion was about to expire, they purchased larger-than-normal quantities, followed by abnormally low purchases in the subsequent month. As a result, in this example, another 6 months were atypical. The process produced outliers in alternating directions in subsequent months.

3. DEALING WITH OUTLIERS

Two approaches to dealing with outliers are possible: (1) automated data modification and (2) manual intervention. Filtering is one method of automated data modification.

In filtering, data points that exceed a predetermined number of standard deviations from the expected datum point are defined as outliers, and the value of that point is changed. Frequently, a 95% confidence interval criterion is established, and any value more than two standard deviations from the value expected interval. In another type of automated data modification, the model uses a weighting process that reduces the impact of the outlier. For example, in regression the regression coefficients can be estimated with an absolute distance algorithm rather than a least square algorithm.

The second method of dealing with outliers is to replace actual data with forecast data for the outlier period. When a forecasting model accurately reflects the underlying process that generates the time series, the value generated by the forecast for the atypical month may provide a very good estimate of what would have occurred without the atypical event. The model is frequently reestimated with the manual intervention technique. Also, the outliers are determined by their distance from the expected value and by the events that generated them. In such a technique, a value that is 1. 8 standard deviations from the expected value and that is associated with a large price cut would be considered an outlier, and the forecast value (the expected value) would replace the actual value for future forecasts and model estimation.

4. AUTOMATED DATA MODIFICATION

There are four approaches to data modification. First is the filtering process, which has been discussed. An alternative approach is a Bayesian analysis of effects of each atypical event.

Let y^*_{t-T} be the datum of the process value of a time series that is T time periods behind t. It is assumed that the forecaster knows that some outside influence was in effect during the period $t + T$, but the exact magnitude of the effect or impulse on the time series is unknown.

4.1 The Multiplicative Correction Factor

The forecaster conditions y^*_{t+T} to reflect uncertainty about the impulse by multiplying y^*_{t+T} by the proportionate change, f_t, attributed to the impulse during the period t + T. Then the conditioned datum is

$$y'_t = (1 + f_{t-T})y^*_{t-T} \tag{1}$$

If the impact of the impulse is known for certain, f_t must be regarded as a random variable representing the various assumed likely levels of proportionate change. Then one would replace f_{t-T} in (1) by its expected value, $E(f_{+T})$, obtaining as the conditioned forecast

$$y'_{t-T} = [1 + E(f_{t-T})]y^*_{t-T} \tag{2}$$

The random variable f_{t-T} can be discrete or continuous. If it is assumed to be discrete, the forecaster must list the set of possible values of f_{t-T} and assign probabilities of their occurrence. However, assuming that f_{t+T} is discrete frequently complicates the analysis and leads to unrealistic results. If it is thought to assume many possible values, it is difficult to assess individual probabilities effectively and hence internal consistency among the probabilities becomes impossible to monitor. Furthermore, it may not be realistic to assume that f_{t+T} will be one of the preselected, discretely spaced values [12].

An alternative approach is to assign a continuous, cumulative probability distribution over a range of values of f_{t+T}. If f_{t+T} is assumed continuous, one would plot the cumulative probability against various values within the range of f_{t+T}; this process would be repeated for different values of f_{t-T} until one is able to construct a smooth, continuous curve through the plot of assessed points.

Although the technique's success is dependent on the precise development of a probability distribution on f_{t+T}, this dependency should not decrease its usefulness. The development of more precise probability distributions is a function of experience with the technique, probability assessments, and the effects of atypical situations.

One valuable method for increasing precision of probability distributions is keeping a journal of atypical events and their effects. The journal should include such information as the causes of a situation, when it was first recognized that the situation would arise, alternative actions available, how long the situation lasted, its effects, and the action taken. The journal will be an aid in assessing effects of subsequent situations.

4.2 The Grouped Approximation Approach

The assessment of a continuous probability distribution on f_{t+T} is not an end in itself in the problem at hand. In order to estimate the expected proportionate gain, $E(f_{t+T})$, one must first discretize the continuous probability distribution, as with the grouped approximation approach.

This approach first identifies a set of 10 equally likely representative values from the probability distribution. The cumulative probability axis is divided into 10 mutually exclusive, collectively exhaustive intervals of equal width; then the median of each interval is found. Thus the representative values are the medians of the 10 equally likely intervals—the values of f_{t+T} associated with the 10 values .05, .15, .25, ..., .95 on the cumulative probability axis. The analysis then proceeds as if the 10 values constitute a collectively exhaustive set of values of f_{t+T} each with an associated probability of .10.

If the ith representative value of the random variable f_{t+T} is denoted as $f_{t+T,j}$, then the expected value of the discretized distribution is found by computing

$$E(f_{t+T}) = (.1)\sum_{i=1}^{10} f_{t+T,i} \tag{3}$$

The term $E(f_{t+T})$ is not necessarily equal to the true expected value of f_{t+T}. However, the approximation should be sufficiently close to allow use of $E(f_{t+T})$ as computed in (3).

4.3 The Incorporation of Sample Information

In many instances, empirical information about the impact of an atypical situation may be available. For example, regular buyers from a wholesaler could be canvassed to assess their behavior if a price-cutting program was not initiated; buyers of domestically manufactured radio and television equipment could be asked if they would have purchased foreign-made equipment in the absence of an import surcharge. Such sample information can be used in conjunction with the management's subjective probability assessment of proportionate change to give an expected change reflecting both management's expectations and buyers' intentions.

Subjective assessments are updated with Bayes' law to reflect sample information. As applied to this problem, if $f_{t+T,k}$ represents the kth representative value from the subjective distribution of proportionate change, then the posterior likelihood of $f_{t+T,k}$ in light of the sample information $\overline{x}$ is

$$P(f_{t-T,k} \mid x) = \frac{P(f_{t-T,k})g^*(z_k)}{\sum_{i=1}^{10} P(f_{t+T,i})g^*(z_i)}$$

where g (z_i) is the Gaussian density function evaluated at $z_i = [n(x - f_{t+T,i})/s_x$, n is the number of customers supplying sample information, $\bar{x}$ is the mean proportionate change obtained from the customer interviews, and s_x is the standard deviation of customer responses.

In the above situations, the sales time series is really two time series. One is a sales demanded time series, and the other is a production capacity sales time series. The demand is what should be forecast. To do this, the forecaster must identify which periods were constrained by supply and increase the sales of those periods to reflect the quantity of product demanded. For a manufacturer, this could be accomplished by adding shorted orders for the period to the amount shipped for the period. Where this cannot be done, the forecaster may have to make an estimate of what the additional sales would have been if there was an adequate supply of product.

An example of improving forecasting accuracy by identifying outliers and using a manual intervention method to deal with outliers is discussed below, using data on tourists going to Hawaii.

The tourist-to-Hawaii time series was used by Geurts and Ibrahim (1975) in a forecasting effort. Bunn (1979) also examined this time series and suggested methods for increasing its accuracy, and this effort was cited by Moriasty and Adams (1979) in a forecasting effort. Data used in the Geurts and Ibrahim study came from the period 1952 to 1972. They included 252 data points; however, only 24 periods were forecast. The time series was forecast by a large number of time series models. The models producing the most accurate forecasts were the $(0,1,1)(1,1,1)_{12}$ Box-Jenkins model and the double smoothed, $\alpha = .1$, exponential smoothing model. The average forecasting error in this project was slightly under 10%.

The time series is very unstable and difficult to forecast. Trend and seasonal factors have changed over time. The Hawaii tourist bureau and several airline companies continue in an effort to modify the seasonal patterns through heavy off-season promotion. In addition, many atypical events have occurred that mask the underlying process by creating outliers. When the manual intervention process previously discussed is used to deal with the outliers, the forecasting accuracy of the time series can be increased. The average for the exponential smoothing forecasting model can be reduced from 10% to 7.5%.

To illustrate the accuracy improvement with manual intervention, the same 24 months were forecast using the same exponential model. However, the outlier was replaced by the forecast values and each January new forecasting parameters were estimated using the modified data. Since more data are now available, forecasts were continued through January 1977. This procedure provided a second comparison over the longer 1965-1977 period.

4.3.1 Atypical Periods

In the tourists-to-Hawaii time series, nine events affecting 16 of 91 periods from May 1969 to January 1977 could be classified as atypical. Table 1 shows the atypical months and the events that caused them.

TABLE 1 Atypical Periods: 1969-1978

Period	Event
March 1970	747 flights start.
September 1970 October 1970 November 1970	Island-wide hotel strike.
December 1970	Pollution on Waikiki Beach. National coverage.
January 1971	Bad mainland weather. Flights canceled.
December 1971 January 1972 February 1972 March 1972	Nonaffinity group fare in effect. West Coast fares reduced 35% ($70). Fare later eliminated (March 1972).
January 1973 February 1973	U.S. changes to Monday holiday for Washington's birthday. Hotels overbook. State institutes operation ALOHA. Tourists placed in private homes, sent home, or slept on cots in hotel lobbies.
August 1973	American Legion convention.
June 1975	Unknown.
November 1975 December 1975	American Medical Association convention.
June 1976	Lions convention.

5. FORECASTING

Since the atypical months have been identified, the time series can be modified and forecast. Two different time frames are used for forecasting. The first covers the period covered in the Geurts and Ibrahim (1975) study. The second extends the time series through 1976. The time series in the earlier study covered the period from June 1969 to May 1971 and was shown to be equally well forecast by a Box-Jenkins $(0,1,1)(1,1,1)_{12}$ model or an exponential double smoothing model with a .1 α (Geurts and Ibrahim, 1975).

5.1 Replication of the Short-Time Series

The procedure used in this study was to reestimate parameter values every year using only the past 24 months. Values for months identified as atypical were replaced by the forecast values for those months.

During the 24 months, four atypical periods influenced 9 months, creating 9 outliers. Table 2, reproduced from Geurts and Ibrahim (1975), shows the forecast errors for the Box-Jenkins model and the exponential smoothing model compared with the yearly reinitialization atypical data modified forecast. As a measure of accuracy, Theil's U statistic is used (Bleimel, 1973). The U value drops from .103 to .075. This improvement does not completely show the impact of data modification, since not all of the 24 months were affected by the modification, yet all are included in the calculated U value.

An atypical event distorts not only the month in which it occurs but also often the forecast for the following months. If the model is an exponential smoothing model, the atypical nonrepresentative data are smoothed into the smoothing statistic. Then the smoothing statistic developed with the nonrepresentative data is used to forecast the next month. Table 3 shows the effect on Theil's U of the forecast for each of the 2 months following the atypical event for the atypical periods covered in the previous study. When the atypical data are replaced by the forecast value, the U drops from .226 for the outlier months without modification to .046 for the outlier months after modification, which is a significant improvement in accuracy.

6. FORECASTING OVER A LONGER TIME FRAME

Since several years of data are available, it is useful to compare forecasting performance over a longer time frame. A forecast of the period from 1965 through 1976 was generated. When an exponential-type model

TABLE 2 Comparison of Forecasting Errors for 24-Month Period of Geurts and Ibrahim Study

Time period	Exponential double smoothing $\alpha = .1$	Data modified for outliers, exponential double smoothing $\alpha = .1$	Vox-Jenkins (0,1,1) (1,1,1)
June 1969	-1883	-140	375
July 1969[a]	14	834	-336
Aug. 1969	1251	988	-2093
Sep. 1969	-819	84	-1539
Oct. 1969	447	-812	140
Nov. 1969	72	-807	644
Dec. 1969	799	226	628
Jan. 1970	162	387	-822
Feb. 1970	1902	-1572	-946
Mar. 1970[a]	-30	428	1247
Apr. 1970[a]	1074	871	275
May 1970	-427	-658	-1454
June 1970	-2629	1063	-692
July 1970	-1697	1573	1263
Aug. 1970	-2545	1880	4269
Sep. 1970[a]	-39	696	49
Oct. 1970[a]	914	1930	-969
Nov. 1970	1331	5757	-1241
Dec. 1970	-665	1	1241
Jan. 1971[a]	2478	-1561	-2179
Feb. 1971[a]	3307	-152	-2381
Mar. 1971[a]	2764	1689	-3861
Apr. 1971	57	-37	-160
May 1971	-781	-889	602
Theil's U	.103	.075	.102

[a] Outlier.

TABLE 3 Subsequent Month Error Changes for Forecasts Where the Outlier Month Data Are Replaced with Previous Forecast Value

Atypical event and date	Error without modification	Date	Error with modification	Actual sales
Operation Aloha and Monday holiday change, Feb./March 1973	9658	April	-1697	20,701
	798	May	-337	19,676
Extraordinary bad weather, airports snowed in, Jan. 1971	3307	Feb.	359	12,672
	2704	March	72	13,630
Hotel strike, Nov. 1970	-665	Dec.	711	15,331
	2478	Jan.	746	11,466
Jumbo jet service starts, March 1970	1074	April	-402	12,472
	427	May	-1	13,069
Total U value for above months	.226		.046	

with an α of .1 was used, the U value obtained was .098, slightly lower than the value for the 24-month forecast in the previous study. However, when the data are modified for outlier months and reinitialized yearly, the forecasting is greatly improved, with a U of .048.

An additional automated approach to dealing with outliers in forecasting is to use a regression equation that is estimated by least lines (absolute distance). Another approach is to estimate the equation with a robust regression which uses least squares for all estimating points except outliers and then to use least lines for the outliers.

The Hawaii tourist data were used to compare the three approaches: (1) least squares, (2) least lines, and (3) robust. The regression equation was estimated using all but the last 12 data points and was then used to forecast the last 12 data points. The data were seasonalized for the three regression estimates and the forecasts were adjusted for the seasonality.

TABLE 4

	Least squares	Robust forecasts	Least lines	Actual
[1,]	363140.2	364129.4	364847.9	483115
[2,]	381738.2	382783.1	383561.0	453811
[3,]	409015.0	410140.0	410997.5	475966
[4,]	378147.2	379192.3	380007.2	456473
[5,]	399692.5	400802.4	401686.9	371028
[6,]	519673.2	521123.0	522303.0	303933
[7,]	517619.3	529098.1	530326.3	366063
[8,]	553577.2	555135.9	556455.9	360670
[9,]	391558.8	392666.2	393622.0	376376
[10,]	408512.7	409673.3	410693.3	383762
[11,]	351243.8	352246.1	353142.7	461832
[12,]	407552.6	408720.7	409783.4	472119
Constant	69985	68895	63369	
6	1511	1520	1548	
		Errors		
	[,1]	[,2]	[,3]	
[1,]	58659.81	57670.62	56952.16	
[2,]	33261.81	32216.88	31438.97	
[3,]	55665.03	54540.03	53682.50	
[4,]	26992.81	25947.72	25132.81	
[5,]	-48252.44	-49362.34	-50246.91	
[6,]	-146603.1	-147152.9	-149332.9	
[7,]	-73159.32	-74638.07	-75866.25	
[8,]	-85447.19	-87005.88	-88325.88	
[9,]	-47238.75	-48346.19	-49301.97	
[10,]	-43522.72	-44683.28	-45703.31	
[11,]	25096.19	24093.88	23197.34	
[12,]	37287.41	36119.34	35056.59	

Table 4 shows the forecast for the three models and also shows the errors for each model. The MAPE was .1406 for the least squares forecast, .1410 for the robust regression forecast, and .1413 for the least lines regression. This is not a statistically significant difference between the errors for the three techniques, and the accuracy level is about half that of the forecast from the manual intervention approach. The conclusion is that if there are frequent outliers in the data to be forecast, the method of dealing with outliers that produces the most accurate forecasts is a manual intervention technique of replacing the outliers with the forecast value for that period and then reestimating the model at frequent intervals.

7. REGRESSION-BASED FORECASTING MODELS

Two commonly used time series techniques are Box-Jenkins (ARIMA) models and exponential smoothing models. Both techniques are regression-type forecasting models. The basic single smoothed exponential model is given by Brown (1963) as

$$\hat{x}_{t+1} = s^{[1]}x_t$$

where

$$s_t^{[1]} = \alpha x_t + [1 - \alpha] s^{[1]} x_{t-1}$$

The weight each prior period has in the single smoothing statistic $s^{[1]}x_t$ is determined by the formula

$$\alpha\beta^k$$

where $\beta = (1 - \alpha)$ and k is the age of the data.

Given the above formula, a single smooth forecasting model can be expressed as a regression equation. For example, if $\alpha = .3$, the single smooth exponential smoothing model could be written as a regression model of the form

$$x_{t+1} = .3x_t + .21x_{t-1} + .14x_{t-2} + .10x_{t-3} + .07x_{t-4} + .05x_{t-5}$$
$$+ .03x_{t-6} + .02x_{t-7} + .02x_{t-8} + .01x_{t-9}$$

8. MARKETING-INDUCED OUTLIERS

Outliers occur frequently in sales time series as the result of marketing efforts. One of the authors of this paper has worked with sales time series in which nearly one in three observations was subject to an atypical event which generated an outlier. For example, in one forecasting effort, the authors encountered a company sales pattern in which the following atypical events occurred over a 26-month period.

1. In July and August, there was a price promotion to wholesalers and large retailers of 10 cents a case.
2. In October, November, and December there was a price promotion of 10 cents a case.
3. In June and July, there was a price promotion of 10 cents a case.
4. The Food and Drug Administration found fungus in a food product. This incident received national coverage in newspapers, resulting in a recall of the product and reduction in sales.
5. A competitive firm increased its price by 20%.
6. The company increased its price 2 months later by 20%.
7. A plant was closed by government inspectors.
8. A competitive price promotion occurred.

In this situation, at least 12 of 26 months could be called outliers. Also, research indicated that the retailers reacted to the price promotion by altering the time when they purchased as well as the quantity purchased. Retailers who were notified of the promotion 6 weeks before the promotion period postponed purchases.

9. OUTLIERS

Outliers are data points that do not fit with the pattern of the data. The value is very low or high in relation to the preceding or following values. Outliers are caused by atypical events, by errors in reporting and recording the data, or by the random make of the data. Handling outliers is a difficult process in any type of data analysis, but in forecasting it is a critical function since the past data, including the outliers, are the basis for forecasting. The dilemma is threefold: (1) if the outlier is either a measurement error or a nonrepresentative part of the underlying process, its inclusion distorts the model's perception of past data and the underlying process; (2) if outliers are removed, the amount of data available is reduced; (3) if the data are modified so that they no longer fall into the outlier classification, then the model creates forecasts generated or based on nonactual data.

REFERENCES

Bleimel, F. (1973). Theil's forecasting accuracy coefficient: A clarification. J. Market. Res. 10 (November): 444-446.

Box, G. E. P., and Jenkins, G. M. (1976). Time Series Analysis Forecasting and Control, rev. ed. Holden-Day, San Francisco.

Brown, R. G. (1963). Smoothing, Forecasting and Prediction of Discrete Time Series. Prentice-Hall, Englewood Cliffs, N.J.

Bunn, D. W. (1979). The synthesis of predictive models in marketing research. J. Market. Res. 116 (May): 280-283.

Fisher, F. M. (1966). The Identification Problem in Econometrics. McGraw-Hill, New York.

Geurts, M. D., and Ibrahim, I. B. (1975). Comparing the Box-Jenkins approach with the exponentially smoothed forecasting model application to Hawaii tourists. J. Market. Res. 12 (May): 182-188.

Geurts, M. D. (1982). Forecasting the Hawaiian tourist market. J. Travel Res. 21(1), Summer.

Krueger, R. (1980). Seasonal adjustment of irregular time series: U.S. merchandise trade. Presented at the third International Time Series Meeting, Houston, August.

Makridakis, S., Wheelwright, S. C., and McGee, V. E. (1983). Forecasting Methods and Applications, 2d ed. Wiley, New York.

Tukey, J. W. (1977). Exploratory Data Analysis. Addison-Wesley, Reading, Mass.

Winters, P. R. (1960). Forecasting sales by exponentially weighted moving averages. Manage. Sci. 6 (April): 324-342.

Part IV

Robust Ridge Regression

13

A Comparison of Regression Estimators When Both Multicollinearity and Outliers Are Present

ROGER C. PFAFFENBERGER and TERRY E. DIELMAN
Texas Christian University, Fort Worth, Texas

1. INTRODUCTION

Two problems which often plague researchers using regression techniques are multicollinearity and nonnormal error distributions.

Multicollinearity is the term used to describe cases in which explanatory variables are correlated among themselves. The effects of multicollinearity are well known and can be found in most textbooks on regression. For example, Neter et al. (1985, pp. 382-400) note that correlation among the explanatory variables may result in imprecise information being available about the population regression coefficients. Adding or deleting variables may change the estimated regression coefficient values, coefficients may have signs that are opposite of what would be expected, and important variables may have coefficients which are statistically insignificant because of inflated standard errors when traditional hypothesis testing procedures are used.

Various remedial measures have been suggested for the multicollinearity problem. One such remedial measure is ridge regression. The ridge regression estimator of the regression coefficients is biased but has smaller sampling variability. Thus the ridge estimator may be preferred to the ordinary least squares estimator if the gain in precision is not offset by the bias. The ridge estimator will be discussed in more detail in Section 2.

The ordinary least squares estimator of the regression coefficients is known to possess certain optimal properties when the disturbances of the regression equation are independent, identically distributed, normal random variables. Specifically, the least squares estimator possesses minimum variance among all unbiased estimators. It is also the maximum likelihood estimator. See, for example, Judge et al. (1985, pp. 11-17).

If the disturbances are not normally distributed, the least squares estimator can be shown to possess minimum variance among all linear unbiased estimators provided the variance of the disturbance distribution is finite. This property and the asymptotic (large-sample) justification of conventional hypothesis testing procedures have often been used to justify using ordinary least squares to estimate regression coefficients even when disturbances are known to be nonnormal. However, recent studies have shown that the least squares estimator may produce extremely poor estimates in the presence of nonnormal disturbance distributions. Also, as pointed out by Koenker (1982), the power of conventional tests may be low when disturbances are nonnormal even though the tests have the correct level of significance asymptotically. Koenker also noted that the class of linear estimators is very restrictive and may exclude estimators which outperform least squares in cases of nonnormal disturbances.

An additional consideration is the research, especially in finance and economics, which suggests that disturbances with infinite variance be used to represent certain economic data. This literature is summarized in Fama (1965, 1970). If the disturbance variance is infinite, the least squares estimator will no longer be minimum variance and both the estimator and the usual hypothesis tests will be unreliable since they depend on an estimator for the disturbance variance.

A search for estimators which are not as strongly affected by nonnormal disturbances has resulted. These robust estimation techniques will be further discussed in Section 3.

The primary purpose of this chapter is to examine estimators which are resistant to the combined problems of multicollinearity and nonnormal disturbance distributions. Specifically, can the ridge estimator and some robust estimation technique be combined to produce a robust ridge regression estimator? In Section 4 the possible alternatives for construction of such an estimator are discussed. Section 5 presents the design and results of a Monte Carlo simulation to examine how well such estimators perform, and Section 6 presents some concluding remarks and suggestions for further research.

2. RIDGE REGRESSION ESTIMATORS

When the explanatory variables in a regression equation are highly correlated, the least squares coefficient estimates will be imprecise. The correlation may be pairwise, or one or more variables may be near-linear combinations of other explanatory variables. The term multicollinearity is often used to characterize such a situation. In cases where multicollinearity exists, one or more of the regression coefficient estimates may have large standard errors. Thus, use of the conventional t-test may suggest that the variables associated with such coefficients are unimportant. The researcher is left uncertain as to whether this is the case or whether the test results are being influenced by standard errors which are inflated due to the multicollinearity. For a further discussion of the multicollinearity problem see Vinod and Ullah (1981, pp. 120-131).

The ridge regression estimator was first proposed by Hoerl (1962) and was subsequently brought to broader attention by Hoerl and Kennard (1970a, 1970b). Initially the use of ridge regression techniques was suggested as a way of examining changes in the estimated regression coefficients that would result from small changes in the values of the explanatory variables. In cases of multicollinearity, small changes in the data values can result in large changes in the estimated coefficient values.

The ridge regression technique has subsequently been adopted as a method which may provide improved estimates of the regression coefficients. That is, ridge regression might be useful in providing estimates which are more precise (and therefore more stable) than least squares estimates in the presence of multicollinearity.

The linear regression model can be written

$$Y = X\beta + e \tag{1}$$

where Y is an $N \times 1$ vector of observations on the dependent variable, X is an $N \times p$ matrix of observations on the explanatory variables, β is a $p \times 1$ vector of regression coefficients to be estimated, and e is an $N \times 1$ vector of disturbances. Throughout this chapter it will be assumed that the rank of X is equal to p ($p < N$) and that the disturbances are independent identically distributed random variables with mean $E(e) = 0$ and variance-covariance matrix $E(ee') = \sigma^2 I$.

The ordinary least squares estimator of β can be written

$$\hat{\beta}_{LS} = (X'X)^{-1}X'Y \tag{2}$$

One way to characterize the multicollinearity problem is through examination of the singular values of the matrix X or, equivalently, through examination of the eigenvalues of the p × p matrix X'X. The matrix X can be written as

$$X = H\Lambda G' \tag{3}$$

where H is an N × p orthogonal matrix, G is a p × p orthogonal matrix, and Λ is a p × p diagonal matrix. The diagonal elements of Λ, denoted $\lambda_1, \ldots, \lambda_p$, are the ordered singular values of X with λ_1 the largest and λ_p the smallest singular value. The decomposition in equation (3) is known as the singular value decomposition.

The matrix X'X can be written

$$X'X = G\Lambda'H'H\Lambda G' \tag{4}$$

$$= G\Lambda^2 G \tag{5}$$

since H'H = I. The squared singular values are the diagonal elements of the matrix Λ^2. These values, λ_i^2, are the eigenvalues of the matrix X'X, while the columns of G are the eigenvectors. Note that the least squares estimator can be written as

$$\hat{\beta}_{LS} = G\Lambda^{-1}H'Y \tag{6}$$

For more on the singular value decomposition see Golub and Reinsch (1970).

If exact multicollinearity exists, then at least one of the eigenvalues of X'X (or, equivalently, at least one of the singular values of X) will be equal to zero and the inverse of X'X will not exist. If the degree of multicollinearity is high but not exact, the smallest eigenvalue will be close to zero. To indicate the degree of multicollinearity, the condition number of the matrix X (or X'X) is often used. The condition number, CN, is defined as the ratio of the largest singular value of X to the smallest

$$CN = \lambda_1/\lambda_p \tag{7}$$

When multicollinearity is present CN will be large, since λ_p will be close to zero. No exact guidelines are available to indicate how large CN must be to indicate severe problems resulting from multicollinearity. However, Belsley et al. (1980, pp. 100–104) suggest that weak dependencies may exist for CN around 5 to 10 and moderate to strong dependencies correspond to CN around 30 to 100. These ranges apply for data that have been scaled to unit length.

The ridge regression estimator is

$$\hat{\beta}_{RID} = (X'X + kI)^{-1}X'Y \tag{8}$$

where k is a constant, often called the biasing parameter, and I is the p × p identity matrix. Thus, to obtain the ridge regression estimator a constant k is added to the diagonal elements of the X'X matrix before inverting it. The resulting estimator will be biased but will have smaller sampling variability than the least squares estimator. Thus it may be possible to obtain a biased but more precise estimator of $\hat{\beta}$ by using ridge regression.

Hoerl and Kennard (1970a) show that there always exists a $k > 0$ so that the mean squared error (MSE) of $\hat{\beta}_{RID}$ will be smaller than the MSE of $\hat{\beta}_{LS}$. The MSE is defined as

$$\text{MSE}(\hat{\beta}) = \text{Var}(\hat{\beta}) + \text{Bias}(\hat{\beta})^2 \tag{9}$$

Their result indicates that for a proper choice of k the introduction of bias into the estimator of β will be offset by a reduction in the variance, thus producing an estimator that is superior to $\hat{\beta}_{LS}$ in a mean squared error sense.

The use of the biasing parameter k improves the conditioning of the matrix to be inverted. The CN for the matrix

$$X'X + kI$$

will be

$$\text{CN} = \left(\frac{\lambda_1^2 + k}{\lambda_p^2 + k}\right)^{1/2} \tag{10}$$

and can be shown to decrease as k increases. Adding a value k to the diagonal elements of X'X will decrease the condition number and thus provide a less ill-conditioned matrix to be inverted.

For comparison with equation (6), note that the ridge estimator can be written

$$\hat{\beta}_{RID} = G(\gamma^2 + kI)^{-1}\gamma H'Y \tag{11}$$

Equation (11) further highlights the fact that the eigenvalues, λ_i^2, will be increased by the value of the constant k.

As mentioned previously, there will always be a value k that produces an estimator $\hat{\beta}_{RID}$ that has a smaller MSE than $\hat{\beta}_{LS}$. The proof of this fact, however, assumes that k is nonstochastic. In practice, the optimal value of k will be unknown. In such cases a variety of methods have been suggested to determine an appropriate value for k from the sample data. When this is done, the assumption that k is nonstochastic is violated and the proof of superiority of $\hat{\beta}_{RID}$ no longer holds. This fact has caused some researchers to question the efficacy of ridge regression when k is chosen from the sample data. For a discussion see Judge et al. (1985, pp. 915–930).

A number of methods of determining k from sample data have appeared in the literature. The earliest of these involved the use of a graphical technique called the ridge trace, which is described in Hoerl and Kennard (1970b). Several nongraphical methods of determining k have also been proposed. Many of these are reviewed in an article by Gibbons (1981). She also provides results of a Monte Carlo simulation comparing ridge estimators using the various methods for choosing k. Her results suggest that three algorithms perform well overall:

1. The Hoerl et al. (1975) estimator of k, denoted k_{HKB}.

$$k_{HKB} = \frac{ps^2}{\hat{\beta}'_{LS}\,\hat{\beta}_{LS}} \tag{12}$$

   where

$$s^2 = \frac{(Y - X\hat{\beta}_{LS})'(Y - X\hat{\beta}_{LS})}{n - p} \tag{13}$$

   is the usual estimator of σ^2 used with least squares regression.
2. The Golub et al. (1979) estimator of k, denoted k_{GHW}. Writing k as k = nc, they propose choosing the value of k that minimizes

$$V(c) = \frac{\|[I - A(c)]Y\|^2}{[\text{Trace}\,(I - A(c))]^2} \tag{14}$$

where $A(c) = X(X'X + ncI)^{-1}X'$ and $||\cdot||$ denotes the Euclidean norm. To find k_{GHW}, $V(c)$ is evaluated for a mesh of k values until the minimizing value, k_{GHW}, is found.

3. The estimator of k proposed by Dempster et al. (1977), denoted k_{DSW}. They suggest choosing k so that

$$\sum_{i=1}^{p} \frac{\hat{\gamma}_i^2}{s^2/k + s^2/\lambda_i} = p \tag{15}$$

where $\hat{\gamma} = G'\hat{\beta}$, $\hat{\gamma}_i$ is a component of the vector $\hat{\gamma}$, G is defined in equation (3), and s^2 is the estimator of σ^2 defined in equation (13).

Gibbons (1981) also reviews other simulation studies designed to examine methods of choosing k. She finds their results to be in agreement with hers overall, although Lawless (1978) and Wichern and Churchill (1978) do present some criticism of the HKB estimator.

There have been a number of simulations comparing ridge estimators to least squares or other estimators. Studies suggesting that ridge estimators are superior to least squares and other estimators under certain conditions include Hoerl et al. (1975), Hoerl and Kennard (1976), Dempster et al. (1977), Lawless and Wang (1976), Gunst and Mason (1977), and Precht and Rao (1985). Draper and Van Nostrand (1979) and Pagel (1981) present criticisms of the design used in certain of the simulation studies. Farebrother (1983) and Silvapulle (1985) reexamine these criticisms and suggest that they are ill founded.

Surveys of research on ridge regression include Alldredge and Gilb (1976), Vinod (1978), and Vinod and Ullah (1981, pp. 169–240). More recent research articles include Nordberg (1982), Hemmerle and Carey (1983), Casella (1985), and Lee and Campbell (1985).

3. ROBUST REGRESSION ESTIMATORS

This section will concentrate on a discussion of estimators that are robust to nonnormal disturbance distributions. Such robust regression estimators are intended to be more efficient estimators than least squares when disturbances are nonnormal, while sacrificing a small amount of efficiency when the assumption of normal disturbances is true. The term "nonnormal disturbances" will indicate disturbance distributions which have fatter tails than the normal distribution. These fat-tailed distributions are more prone to produce outliers than is the normal distribution. The adverse effect of outliers on the least-squares-fitted regression line is well known and it is protection against the effect of outliers that is being sought through the use of robust estimators.

Also, throughout this chapter the term outlier will mean an extreme observation in the dependent variable space. An outlier might be produced by a coding error or might be a correct value but one which is unusual when compared with most of the observations. Observations which are unusual in the independent variable space will not be dealt with in this chapter. These observations are usually called influential observations.

Several different classifications of robust methods exist. Three of the most commonly considered groups are R-estimators, L-estimators, and M-estimators. R-estimators are estimators based on ranking of the residuals from the regression. To obtain R-estimates of the regression coefficients, the function to be minimized can be written

$$\sum_{i=1}^{N} a(R_i)\hat{e}_i \qquad (16)$$

where $\hat{e}_i = Y_i - X_i\hat{\beta}$ is the regression residual, R_i is the rank of $\hat{e}_i$, and $a(\cdot)$ is some monotone scores function satisfying

$$\sum_{i=1}^{N} a(i) = 0 \qquad (17)$$

For more detail on R-estimation see Jaeckel (1972) or Jureckova (1971, 1977).

A second class of robust estimator is called the L-estimators. L-estimators are linear combinations of the order statistics. As measures of location the construction of L-estimators is fairly straightforward. For example, the sample quantiles are L-estimators and any linear combination of the sample quantiles would also be an L-estimator.

The extension to the case of linear regression is more complex. However, Koenker and Bassett (1978) developed estimators called the regression quantiles that generalize the L-estimators to linear models. They define the θth regression quantile as the solution to the minimization problem

$$\min_{\beta} \sum_{\{i \mid y_i \geq x_i'\beta\}} \theta \, | y_i - x_i'\beta | + \sum_{\{i \mid y_i < x_i'\beta\}} (1 - \theta) \, | y_i - x_i'\beta | \qquad (18)$$

where $0 < \theta < 1$, y_i is the ith observation on the dependent variable, x_i is the $p \times 1$ vector of independent variable values for the ith observation, and β is the $p \times 1$ vector of unknown regression coefficients to be estimated. Note that when $p = 1$ and the x values are identically one, the solution to the problem in equation (18) reduces to the θth sample quantiles. Also, when $\theta = 1/2$, the minimization problem in equation (18) becomes

$$\min_{\beta} \sum_{i=1}^{N} |y_i - x_i'\beta| \tag{19}$$

or minimizing the sum of the absolute values of the regression residuals. The solution to this problem is called the least absolute value (LAV) estimator and will be discussed in more detail later in this section.

For further reading on regression quantiles see Koenker and Bassett (1982b) and Bassett and Koenker (1982).

The third and final class of robust estimators to be discussed in this section is called the class of M-estimators.

M-estimators for linear regression are defined as solutions to the following minimization problem:

$$\min_{\beta} \sum_{i=1}^{N} \rho(y_i - x_i'\beta) \tag{20}$$

where $\rho(\cdot)$ is a strictly convex function of the regression residuals. Alternatively, M-estimators can be defined as the solution to the equations

$$\sum_{i=1}^{N} x_i \psi(y_i - x_i'\beta) = 0 \tag{21}$$

where the derivative of $\rho(\cdot)$ is the function $\psi(\cdot)$. The terms y_i, x_i, and β are as defined for equation (18).

The least squares estimator can be defined as an M-estimator by defining

$$\rho(y_i - x_i'\beta) = 1/2(y_i - x_i'\beta)^2 \tag{22}$$

and

$$\psi(y_i - x_i'\beta) = y_i - x_i'\beta \tag{23}$$

Since the purpose of robust estimation procedures is to reduce the effect of outliers on parameter estimates, the definition of $\rho(\cdot)$, or equivalently $\psi(\cdot)$, is usually made in some manner intended to achieve this purpose. For example, Huber (1964) proposed the following $\psi(\cdot)$ function:

$$(y_i - x_i'\beta) = \begin{cases} y_i - x_i'\beta & \text{if } |y_i - x_i'\beta| \leq c \\ c \operatorname{sign}(y_i - x_i'\beta) & \text{if } |y_i - x_i'\beta| > c \end{cases} \tag{24}$$

Thus residuals larger in absolute value than some chosen constant c are given less weight in the determination of an estimate of β. Typically, in computing M-estimates the residuals are divided by some robust estimate of scale so that the constant c can be assigned a standardized value such as $c = 2.0$.

Other possible $\psi(\cdot)$ functions have been suggested by Ramsay (1977), Hampel (see Andrews et al., 1972), and Andrews (1974) and are discussed conveniently in Montgomery and Peck (1982, p. 369).

For more on M-estimators see Huber (1973) and Yohai and Maronna (1979).

The LAV estimator was mentioned earlier as one of the class of regression L-estimators. It can also be shown to be a special class of the M-estimators. Since LAV estimation plays a primary role in subsequent sections of this chapter, it will be given special attention at this point. The LAV estimator, $\hat{\beta}_{LAV}$, can be defined as the solution to the problem

$$\min_{\beta} \sum_{i=1}^{N} |y_i - x_i'\beta| \tag{25}$$

Rather than minimizing the sum of squared residuals as in least squares estimation, the sum of the absolute values of the residuals is minimized. Thus, the effect of outliers on the LAV estimates will be less than that on least squares estimates. The LAV estimation problem can be set up as a linear programming problem, as shown by Wagner (1959) and others, as follows:

$$\text{minimize} \sum_{i=1}^{N} (d_i^+ + d_i^-) \tag{26}$$

subject to

$$y_i - x_i'\beta + d_i^+ - d_i^- = 0, \quad i = 1, \ldots, N$$

$$d_i^+, d_i^- \geqslant 0, \quad i = 1, \ldots, N$$

β unrestricted in sign

where d_i^+ and d_i^- are, respectively, the positive and negative residuals associated with the ith observation.

The formulation of the LAV estimation problem as a linear programming problem has led to the development of very fast special-purpose algorithms for producing LAV estimates, including algorithms by Barrodale and Roberts (1973, 1974) and Armstrong et al. (1979).

The asymptotic distribution theory and procedures for inference have also been studied more thoroughly for LAV estimation than for other robust procedures. See Bassett and Koenker (1978), Koenker and Bassett (1982a), Dielman and Pfaffenberger (1982), and Sposito and Tveite (1986).

For further reading on LAV estimation, see the surveys by Dielman (1984), Dielman and Pfaffenberger (1982, 1984), and Narula and Wellington (1982) and the references listed in those papers.

For other surveys on robust estimators for linear regression see Hogg (1979), Huber (1981), and Koenker (1982). Adaptive approaches to robust estimation are discussed by Hogg (1974).

4. ROBUST RIDGE REGRESSION ESTIMATORS

When both outliers and multicollinearity occur in a data set, it would seem beneficial to combine methods designed to deal with these problems individually. This section presents some possible combinations of ridge regression estimation discussed in Section 2 and robust procedures discussed in Section 3. It is hoped that the resulting robust ridge regression estimators will be resistant to multicollinearity problems and less affected by extreme observations.

Askin and Montgomery (1980) discuss augmented robust estimators as a way of combining biased and robust regression techniques. Similar estimators are suggested by Hogg (1979) and Vinod and Ullah (1981, p. 325).

The procedure suggested by Askin and Montgomery is based on the fact that M-estimates can be computed using weighted least squares procedures. The weighted least squares estimator can be written as

$$\hat{\beta}_{WLS} = (X'WX)^{-1}X'WY \tag{27}$$

where W is a diagonal matrix with diagonal elements W_{ii}. The W_{ii} are weights applied to the observations and are intended to downweight the extreme observations. The weights can be determined using any M-estimate ψ function (or any other weight which appropriately downweights extreme observations).

Askin and Montgomery suggest using weights

$$W_{ii} = \frac{\psi(y_i - x_i'\beta)}{y_i - x_i'\beta} \tag{28}$$

The weighted least squares estimates can be computed by applying least squares to the transformed observations $\sqrt{W_{ii}}\, y_i$ and $\sqrt{W_{ii}}\, x_i$. This produces an estimate equivalent to $\hat{\beta}_{WLS}$ in equation (27). Note that the estimation procedure can be iterated to produce what are called the iteratively reweighted least squares estimates. See Montgomery and Peck (1982, p. 368).

To compute robust ridge estimates the formula

$$\hat{\beta}_{WRID} = (X'WX + kI)^{-1}X'WY \tag{29}$$

can be used. Askin and Montgomery compute the value of k from the untransformed observations, although the transformed observations could be used. The estimator $\hat{\beta}_{WRID}$ with the biasing parameter k and the weights W_{ii} determined from the data will be referred to as the weighted ridge (WRID) estimator throughout the remainder of this chapter. See Lawrence and Marsh (1984) for applications of the WRID estimator using various weighting schemes.

Another robust ridge regression estimator was suggested by Pfaffenberger and Dielman (1984, 1985). This estimator combines properties of the LAV estimator and the ridge estimator and will be referred to as the RLAV estimator. The RLAV estimator can be written as

$$\hat{\beta}_{RLAV} = (X'X + k^*I)^{-1}X'Y \tag{30}$$

The value of k^* is determined from the data using

$$k^* = \frac{ps^2_{LAV}}{\hat{\beta}'_{LAV}\hat{\beta}_{LAV}} \tag{31}$$

where

$$s^2_{LAV} = \frac{(Y - X\hat{\beta}_{LAV})'(Y - X\hat{\beta}_{LAV})}{n - p} \tag{32}$$

and $\hat{\beta}_{LAV}$ is the LAV estimator defined as the solution to equation (25). The value k^* is the estimator of k suggested by Hoerl et al. (1975) and described in equation (12), with two important exceptions. First, the LAV estimator of β is used rather than the least squares estimator. Second, the estimator of σ^2 used in formula (31) is modified by using the LAV coefficient estimates rather than the least squares estimates. These two changes are intended to reduce the effect of outliers on the value chosen for the biasing parameter.

5. COMPARISON OF ROBUST RIDGE REGRESSION ESTIMATORS

5.1 Purpose of the Experiment

A Monte Carlo simulation was designed to compare the performance of several estimators. The simulation was designed to allow both multicollinearity and nonnormal disturbance distributions to be present simultaneously. Varying degrees of multicollinearity were allowed. The nonnormal distributions used to generate disturbances were fat-tailed distributions which are prone to produce outliers.

The five estimators included in the study were

1. Least squares
2. Least absolute value (LAV)
3. Ridge regression
4. RLAV
5. WRID

The least squares estimator was defined in equation (2). The ridge regression estimator was defined in equation (8). The value of k was determined from the data using the k_{HKB} value in equation (12). The LAV estimator was defined as the solution to the minimization problem in equation (25). The algorithm of Armstrong et al. (1979) was used to compute the LAV estimates. The RLAV estimator was defined in equation (30) with k^* of equation (31) used to determine a value for k. The WRID estimator was defined in equation (29). The value of k used was the k_{HKB} value determined from the weighted least squares regression. The diagonal elements of the W matrix were set equal to

$$W_{ii} = 1/|\hat{e}_i| \tag{33}$$

where the $\hat{e}_i$ were residuals from an initial LAV fit to the data. The LAV residuals were used rather than the least squares residuals since the least squares fit is likely to be affected by outliers. To avoid division by zero, when $|\hat{e}_i| < \delta$, $|\hat{e}_i|$ was set equal to δ, where δ represents an arbitrary small number. The simulations used $\delta = 0.01$ although other values were tried. The results were similar for the other δ values tried and are not reported here.

5.2 Design of the Experiment

The model used was

$$y_i = \beta_0 + \beta_1 x_{i1} + \beta_2 x_{i2} + e_i \tag{34}$$

The parameter values β_0, β_1, and β_2 were set equal to one. The explanatory variables x_{i1} and x_{i2} were generated as follows.

$$x_{ij} = (1 - \rho^2)Z_{ij} + \rho Z_{i3}, \quad i = 1, \ldots, N; \; j = 1, 2 \tag{35}$$

where the Z_{ij} are independent standard normal random numbers generated by the IMSL subroutine GGNPM. Once generated for a given sample size N, the explanatory variable values were held fixed throughout the experiment. The value of ρ^2 represents the first factor in the experiment. Its values were chosen as 0.0, 0.5, 0.8, 0.9, 0.95, and 0.99. These numbers represent the correlation between the two explanatory variables. The second factor was the sample size N. Sample sizes of 20, 50, and 100 were examined. The final factor was the disturbance distribution. The following four disturbance distributions were examined:

1. Standard normal distribution.
2. Laplace distribution with mean zero and variance two.
3. Cauchy distribution with median zero and scale parameter one.
4. Contaminated normal distribution where e_i is standard normal with probability 0.85 and normal with mean zero and standard deviation five with probability 0.15.

All random numbers used were generated from IMSL subroutines. Standard normal random numbers used the GGNPM subroutine. Laplace deviates were generated from uniform (0,1) random numbers from the subroutine GGUBFS using the inverse C.D.F. transformation. Cauchy disturbances were obtained using GGCAY. Contaminated normal disturbances were generated using uniform (0,1) deviates from GGUBFS to determine the appropriate distribution and GGNPM to generate normal random numbers. The standard normal values were multiplied by five when the contaminating distribution was chosen. In all cases disturbances were generated independently of the explanatory variables.

The simulations were performed on an IBM 4341 Model 12. Programs were written in double-precision FORTRAN.

For each of the 6 × 3 × 4 = 72 treatments in the three-factor experiment, 1000 Monte Carlo trials were used. For each of the 72 treatments, the following statistics were computed for each of the five estimators:

1. The average of the estimates.
2. The mean squared error (MSE) and the 10 pairwise MSE ratios where

$$\text{MSE} = \sum_{i=1}^{1000} \frac{(\hat{\beta}_i - \beta_i)^2}{1000}$$

3. The mean absolute deviations (MAD) and the 10 pairwise MAD ratios where

$$\text{MAD} = \sum_{i=1}^{1000} \frac{|\hat{\beta}_i - \beta_i|}{1000}$$

4. The 10 pairwise comparisons of "closeness" to the actual parameter values. The closeness measure was computed as the number of times estimator A was closer than estimator B to the true parameter value.

5.3 Results

First consider the comparison of the two robust ridge estimators, RLAV and WRID.

Table 1 shows the number of times the RLAV estimates were closer than the WRID estimates to the true value of the parameter β_1. Values for the parameter β_2 were nearly identical and are not included here. The hypothesis of no difference between the estimators can be tested as follows using a 5% level of significance:

H_0: $\pi = .5$

H_1: $\pi \neq .5$

where π is the true proportion of times RLAV was closer than WRID.

The decision rule for the test can be stated as follows:

Reject H_0 (in favor of RLAV) if tabled entry is 531 or more.
Reject H_0 (in favor of WRID) if tabled entry is 469 or less.
Accept H_0 (no difference) if tabled entry is between 469 and 531.

As can be seen, RLAV performs significantly better than WRID over a wide range of combinations of ρ^2 and the error distribution.

These results are supported by the mean squared estimation error ratios in Table 2. These ratios represent the efficiency of RLAV relative

TABLE 1 Number of Times RLAV was Closer Than WRID to the True Parameter Value $(\beta_1)^a$

	Values of ρ^2					
Error distribution	0.0	0.5	0.8	0.9	0.95	0.99
			N = 20			
Normal	595	629	824	938	973	996
Contaminated normal	395	511	795	920	959	982
Laplace	467	543	788	919	965	988
Cauchy	266	394	700	818	893	957
			N = 100			
Normal	653	659	733	865	941	986
Contaminated normal	352	452	670	878	960	981
Laplace	426	441	586	788	924	980
Cauchy	115	174	460	674	784	880

[a] RLAV is significantly better than WRID (at the 5% level of significance) when counts are 531 or more; WRID is significantly better than RLAV when counts are 469 or less.

TABLE 2 Mean Squared Error Ratios of RLAV to WRID for Estimation of β_1[a]

Error distribution	Values of ρ^2 0.0	0.5	0.8	0.9	0.95	0.99
			N = 20			
Normal	0.74	0.55	0.28	0.20	0.18	0.17
Contaminated normal	1.68	0.96	0.29	0.18	0.15	0.14
Laplace	1.00	0.76	0.30	0.18	0.16	0.15
Cauchy	2.14	1.00	0.28	0.17	0.14	0.12
			N = 100			
Normal	0.64	0.55	0.41	0.26	0.19	0.18
Contaminated normal	2.15	1.47	0.50	0.20	0.14	0.12
Laplace	1.34	1.24	0.62	0.28	0.18	0.15
Cauchy	15.63	5.58	0.90	0.24	0.10	0.03

[a] Values less than one indicate RLAV is more efficient than WRID; values greater than one indicate WRID is more efficient than RLAV.

to WRID. Thus, values less than one indicate that RLAV is more efficient, while values greater than one indicate that WRID is more efficient.

The results presented here are for sample sizes of N = 20 and 100. Results for N = 50 were, in general, similar.

To further describe the degree of multicollinearity present, Table 3 provides the condition numbers (CN) of the X matrices for the chosen values of ρ^2.

TABLE 3 Condition Numbers of X Matrices for Values of ρ^2

ρ^2	0.0	0.5	0.8	0.9	0.95	0.99
CN (for N = 20)	1.3	1.9	5.4	11.5	23.6	121.1
CN (for N = 100)	1.2	2.6	7.4	15.6	32.1	163.9

TABLE 4 Number of Times RLAV Was Closer Than RIDGE to the True Parameter Value (β_1)[a]

Error distribution	Values of ρ^2 0.0	0.5	0.8	0.9	0.95	0.99
			N = 20			
Normal	502	512	492	466	448	422
Contaminated normal	540	613	660	655	651	624
Laplace	525	595	627	595	573	555
Cauchy	588	648	736	773	803	809
			N = 100			
Normal	492	493	454	414	425	407
Contaminated normal	609	667	712	701	684	663
Laplace	573	637	627	634	633	619
Cauchy	632	691	742	774	826	859

[a] RLAV is significantly better than RIDGE (at the 5% level of significance) when counts are 531 or more; RIDGE is significantly better than RLAV when counts are 469 or less.

Since the RLAV estimator clearly outperforms the WRID estimator, the remaining comparisons will be restricted to RLAV to conserve space. Tables 4, 6, and 8 show the number of times the RLAV estimates were closer than the RIDGE, LAV, and LS estimates, respectively, to the true value of the parameter β_1. The hypothesis of no difference between RLAV and each of the estimators can be tested using the decision rule previously stated:

Reject H_0 (in favor of RLAV) if tabled entry is 531 or more.
Reject H_0 (in favor of RIDGE, LAV, or LS) if tabled entry is 469 or less
Accept H_0 (no difference) if tabled entry is between 469 and 531.

The MSE ratios of RLAV to each of the estimators RIDGE, LAV, and LS are given in Tables 5, 7, and 9, respectively. Again, values less than one indicate that RLAV is more efficient, while values greater than one indicate that the other estimator is more efficient.

TABLE 5 Mean Squared Error Ratios of RLAV to RIDGE for Estimation of β_1[a]

Error distribution	Values of ρ^2 0.0	0.5	0.8	0.9	0.95	0.99
			$\underline{N} = \underline{20}$			
Normal	1.02	0.99	0.98	1.00	1.03	1.04
Contaminated normal	0.94	0.79	0.49	0.37	0.33	0.31
Laplace	0.98	0.94	0.76	0.67	0.63	0.62
Cauchy	0.007	0.003	0.003	0.003	0.003	0.003
			$\underline{N} = \underline{100}$			
Normal	1.00	1.00	1.00	1.02	1.04	1.07
Contaminated normal	0.97	0.93	0.59	0.33	0.25	0.22
Laplace	0.99	0.99	0.86	0.63	0.50	0.47
Cauchy	0.011	0.025	0.000	0.000	0.000	0.000

[a] Values less than one indicate RLAV is more efficient than RIDGE; values greater than one indicate RIDGE is more efficient than RLAV.

TABLE 6 Number of Times RLAV Was Closer Than LAV to the True Parameter Value (β_1)[a]

Error distribution	Values of ρ^2 0.0	0.5	0.8	0.9	0.95	0.99
			$\underline{N} = \underline{20}$			
Normal	596	634	823	938	973	998
Contaminated normal	393	515	792	920	964	993
Laplace	471	545	788	924	968	992
Cauchy	268	398	708	836	918	989
			$\underline{N} = \underline{100}$			
Normal	651	661	737	869	952	996
Contaminated normal	350	450	672	884	965	995
Laplace	425	441	589	791	926	991
Cauchy	115	177	471	717	843	972

[a] RLAV is significantly better than LAV (at the 5% level of significance) when counts are 531 or more; LAV is significantly better than RLAV when counts are 469 or less.

TABLE 7 Mean Squared Error Ratios of RLAV to LAV for Estimation of β_1[a]

	Values of ρ^2					
Error distribution	0.0	0.5	0.8	0.9	0.95	0.99
			N = 20			
Normal	0.73	0.55	0.28	0.20	0.17	0.17
Contaminated normal	1.67	0.95	0.29	0.18	0.15	0.14
Laplace	1.00	0.76	0.30	0.18	0.15	0.15
Cauchy	2.11	0.99	0.28	0.16	0.13	0.11
			N = 100			
Normal	0.63	0.55	0.40	0.25	0.19	0.17
Contaminated normal	2.14	1.46	0.50	0.20	0.13	0.11
Laplace	1.33	1.23	0.62	0.27	0.17	0.14
Cauchy	15.5	5.48	0.84	0.22	0.08	0.02

[a] Values less than one indicate RLAV is more efficient than LAV; values greater than one indicate LAV is more efficient than RLAV.

From Tables 4 and 5, RIDGE marginally outperforms RLAV when disturbances are normal and the correlation is high. Otherwise RLAV is superior.

From Tables 6 and 7, LAV outperforms RLAV when the correlation is zero or low and disturbances are nonnormal. Otherwise RLAV is superior.

From Tables 8 and 9, LS is superior when there is no correlation (except for Cauchy disturbances). Otherwise RLAV is superior.

The results from the comparisons of RLAV to the RIDGE, LAV, and LS estimators are not entirely unexpected, given the properties of the various estimators. The two most interesting results from these comparisons are that

1. Using RLAV rather than RIDGE results in a very small loss in efficiency in cases where RIDGE is expected to perform well.
2. RLAV is superior to LAV over a wide range of values of ρ^2 for the given disturbance distributions.

TABLE 8 Number of Times RLAV Was Closer Than LS to the True Parameter Value (β_1)[a]

	Values of ρ^2					
Error distribution	0.0	0.5	0.8	0.9	0.95	0.99
			$\underline{N} = \underline{20}$			
Normal	430	602	912	950	974	995
Contaminated normal	403	662	909	959	985	993
Laplace	429	641	916	950	967	995
Cauchy	615	801	932	968	984	998
			$\underline{N} = \underline{100}$			
Normal	446	691	944	969	983	996
Contaminated normal	470	774	946	968	982	998
Laplace	467	729	951	977	985	998
Cauchy	685	811	924	958	977	997

[a] RLAV is significantly better than LS (at the 5% level of significance) when counts are 531 or more; LS is significantly better than RLAV when counts are 469 or less.

TABLE 9 Mean Squared Error Ratios of RLAV to LS for Estimation of β_1[a]

	Values of ρ^2					
Error distribution	0.0	0.5	0.8	0.9	0.95	0.99
			$\underline{N} = \underline{20}$			
Normal	1.08	0.82	0.44	0.30	0.27	0.26
Contaminated normal	0.93	0.46	0.15	0.09	0.07	0.07
Laplace	1.03	0.69	0.27	0.17	0.15	0.14
Cauchy	0.003	0.001	0.001	0.000	0.000	0.000
			$\underline{N} = \underline{100}$			
Normal	1.00	0.92	0.64	0.40	0.30	0.26
Contaminated normal	0.93	0.68	0.22	0.09	0.06	0.05
Laplace	0.97	0.84	0.42	0.19	0.12	0.10
Cauchy	0.003	0.001	0.000	0.000	0.000	0.000

[a] Values less than one indicate RLAV is more efficient than LS; values greater than one indicate LS is more efficient than RLAV.

TABLE 10 Number of Times RLAV Was Closer Than WRID to the True Parameter Value (β_1) When LAV k Value Is Used

Error distribution	Values of ρ^2					
	0.0	0.5	0.8	0.9	0.95	0.99
			$\underline{N} = \underline{20}$			
Normal	595	635	815	925	953	964
Contaminated normal	389	496	732	833	873	911
Laplace	465	541	766	874	901	941
Cauchy	266	310	458	537	567	639
			$\underline{N} = \underline{100}$			
Normal	653	661	721	835	914	948
Contaminated normal	353	443	602	750	813	865
Laplace	425	433	559	712	833	880
Cauchy	181	182	243	262	291	389

[a] RLAV is significantly better than WRID (at the 5% level of significance) when counts are 531 or more; WRID is significantly better than RLAV when counts are 469 or less.

TABLE 11 Mean Squared Error Ratios of RLAV to WRID for Estimation of β_1 When the LAV k Value Is Used[a]

Error distribution	Values of ρ^2					
	0.0	0.5	0.8	0.9	0.95	0.99
			$\underline{N} = \underline{20}$			
Normal	0.74	0.56	0.31	0.23	0.20	0.20
Contaminated normal	1.67	1.03	0.37	0.25	0.22	0.20
Laplace	1.01	0.80	0.35	0.23	0.20	0.20
Cauchy	1.97	1.37	0.61	0.41	0.34	0.31
			$\underline{N} = \underline{100}$			
Normal	0.64	0.55	0.43	0.30	0.24	0.22
Contaminated normal	2.17	1.55	0.66	0.33	0.25	0.23
Laplace	1.34	1.26	0.70	0.36	0.26	0.24
Cauchy	2.71	2.55	2.20	1.77	1.15	0.48

[a] Values less than one indicate RLAV is more efficient than WRID; values greater than one indicate WRID is more efficient than RLAV.

An additional issue involved in the computation of the WRID estimator is how to compute the value of k. Two possible alternatives are (1) use the value of k computed from the LAV regression used to determine the weights as in equation (31) or (2) after weighting the original observations use an LS regression to determine the value of k as in equation (12).

For the results in Tables 1 and 2, the value of k was computed in the latter manner, by recomputing k from the weighted observations. Tables 10 and 11 show the results when the former scheme is used, computing the k value from the LAV regression. Otherwise the WRID procedure is the same as previously described. The WRID procedure seems to perform slightly better when the LAV k value is used. However, RLAV still appears superior except for low values of ρ^2.

6. CONCLUDING REMARKS

The comparison of the two robust ridge estimators indicates that RLAV is superior to WRID for many combinations of error distribution type and degree of multicollinearity. The RLAV estimator is less efficient than the RIDGE estimator when disturbances are normal. However, the loss in efficiency is small (7% at most in this experiment) and the efficiency gains are dramatic when disturbances are nonnormal. RLAV outperforms both LAV and LS when the degree of multicollinearity is high. Thus, the RLAV estimator appears to be a viable alternative to other estimators when both multicollinearity and nonnormal disturbances are present.

There are limitations to the current study, however. First, since this is a simulation study, its limitations must be recognized. Data have been generated in a manner to try and allow generalization to practical situations, however.

Second, one specific form of the WRID estimator was compared to the RLAV estimator. Many other possible weighting schemes could be used to construct the WRID estimator. Some of these are discussed in Lawrence and Marsh (1984) and Askin and Montgomery (1980). Also, Askin and Montgomery (1984) compared several robust ridge estimators. Their results indicate that the WRID estimator using a weighting scheme due to Hampel performs very well.

Our initial results indicate that the WRID estimator with Hampel weighting performs well relative to the RLAV estimator. More complete comparison of this and other forms of the WRID estimator to RLAV is one avenue of research under current investigation.

ACKNOWLEDGMENT

The contribution of the second author to this research was supported by a grant from the M. J. Neeley School of Business, Texas Christian University.

REFERENCES

Alldredge, J. R., and Gilb, N. S. (1976). Ridge regression: An annotated bibliography. Int. Stat. Rev. 44: 355-360.

Andrews, D. F. (1974). A robust method for multiple linear regression. Technometrics 16: 523-531.

Andrews, D. F., Bickel, P. J., Hampel, F. R., Huber, P. J., Rogers, W. H., and Tukey, J. W. (1972). Robust Estimates of Location. Princeton University Press, Princeton, N. J.

Armstrong, R. D., Frome, E., and Kung, D. (1979). A revised simplex algorithm for the absolute deviation curve-fitting problem. Commun. Stat. B Simul. Comput. 8: 175-190.

Askin, R. G., and Montgomery, D. C. (1984). An analysis of constrained robust regression estimators. Naval Res. Logist. Quart. 31: 283-296.

Askin, R. G., and Montgomery, D. C. (1980). Augmented robust estimators. Technometrics 22: 333-341.

Barrodale, I., and Roberts, F. (1974). Solution of an over-determined system of equations in the L_1 norm. Commun. Assoc. Comput. Mach. 17: 319-320.

Barrodale, I., and Roberts, F. (1973). An improved algorithm for discrete L_1 linear approximation. SIAM J. Numer. Anal. 10: 839-848.

Bassett, G. W., and Koenker, R. W. (1982). An empirical quantile function for linear models with iid errors. J. Am. Stat. Assoc. 77: 407-415.

Bassett, G., and Koenker, R. (1978). Asymptotic theory of least absolute error regressions. J. Am. Stat. Assoc. 73: 618-622.

Belsley, D., Kuh, E., and Welsh, R. E. (1980). Regression Diagnostics. Wiley, New York.

Casella, G. (1985). Condition numbers and minimax ridge-regression estimators. J. Am. Stat. Assoc. 80: 753-758.

Dempster, A. P., Schatzoff, M., and Wermuth, N. (1977). A simulation study of alternatives to ordinary least squares. J. Am. Stat. Assoc. 72: 77-106.

Dielman, T. E. (1984). Least absolute value estimation in regression models: An annotated bibliography. Commun. Stat. A Theory Methods 13: 513-541.

Dielman, T. E., and Pfaffenberger, R. C. (1984). Computational algorithms for calculating least absolute value and Chebyshev estimates for multiple regression. Am. J. Math. Manage. Sci. 4: 169-197.

Dielman, T. E., and Pfaffenberger, R. C. (1982). LAV (least absolute value) estimation in linear regression: A review. TIMS Stud. Manage. Sci.: Optimization Stat. 19: 31-52.

Draper, N. R., and Van Nostrand, R. C. (1979). Ridge regression and James-Stein estimation: Review and comments. Technometrics 21: 451-466.

Fama, E. F. (1970). Efficient capital markets: A review of theory and empirical work. J. Finance 25: 383-417.

Fama, E. F. (1965). The behavior of stock market prices. J. Business 38: 34-105.

Farebrother, R. W. (1983). An examination of recent criticisms of ridge regression simulation designs. Commun. Stat. A Theory Methods 12: 2549-2555.

Gibbons, D. (1981). A simulation study of some ridge estimators. J. Am. Stat. Assoc. 76: 131-139.

Golub, G. H., and Reinsch, C. (1970). Singular value decomposition and least square solutions. Numer. Math. 14: 403-420.

Golub, G. H., Heath, M., and Wahba, G. (1979). Generalized cross-validation as a method for choosing a good ridge parameter. Technometrics 21: 215-223.

Gunst, R. F., and Mason, R. L. (1977). Biased estimation in regression: An evaluation using mean squared error. J. Am. Stat. Assoc. 72: 616-628.

Hemmerle, W. J., and Carey, M. B. (1983). Some properties of generalized ridge estimators. Commun. Stat. B Simul. Comput. 12: 239-253.

Hoerl, A. E. (1962). Application of ridge analysis to regression problems. Chem. Eng. Progress 58: 54-59.

Hoerl, A. E., and Kennard, R. W. (1976). Ridge regression: Iterative estimation of the biasing parameter. Commun. Stat. A Theory Methods 5: 77-88.

Hoerl, A. E., and Kennard, R. W. (1970a). Ridge regression: Biased estimation for nonorthogonal problems. Technometrics 12: 55-67.

Hoerl, A. E., and Kennard, R. W. (1970b). Ridge regression: Applications to nonorthogonal problems. Technometrics 12: 69-82.

Hoerl, A. E., Kennard, R. W., and Baldwin, K. F. (1975). Ridge regression: Some simulations. Commun. Stat. 4: 105-123.

Hogg, R. V. (1979). An introduction to robust estimation. Robustness in Statistics, R. L. Lanner and G. N. Wilkinson, eds., pp. 1-17. Academic Press, New York.

Hogg, R. V. (1974). Adaptive robust procedures: A partial review and some suggestions for future applications and theory. J. Am. Stat. Assoc. 69: 909-923 (with comments, 923-927).

Huber, P. J. (1981). Robust Statistics. Wiley, New York.

Huber, P. J. (1973). Robust regression: Asymptotics, conjectures and Monte Carlo. Ann. Stat. 1: 799-821.

Huber, P. J. (1964). Robust estimation of a location parameter. Ann. Math. Stat. 35: 73-101.

Jaeckel, L. A. (1972). Estimating regression coefficients by minimizing the dispersion of the residuals. Ann. Math. Stat. 43: 1449-1458.

Judge, G. G., and Bock, M. (1983). Biased estimation. Handbook of Econometrics, Vol. 1, Z. Griliches and M. Intriligator, eds., pp. 599-649. North-Holland, Amsterdam.

Judge, G. G., Griffiths, W. E., Hill, R. C., Lutkepohl, H., and Lee, T. (1985). The Theory and Practice of Econometrics. Wiley, New York.

Jureckova, J. (1977). Asymptotic relations of M-estimates and R-estimates in linear regression models. Ann. Stat. 5: 464-472.

Jureckova, J. (1971). Nonparametric estimates of regression coefficients. Ann. Math. Stat. 42: 1328-1338.

Koenker, R. W. (1982). Robust methods in econometrics. Econometric Rev. 1: 213-290.

Koenker, R., and Bassett, G. (1982a). Tests of linear hypotheses and L_1 estimation. Econometrica 50: 1577-1583.

Koenker, R. W., and Bassett, G. W. (1982b). Robust tests for heteroscedasticity based on regression quantiles. Econometrica 50: 43-62.

Koenker, R. W., and Bassett, G. W. (1978). Regression quantiles. Econometrica 46: 33-50.

Lawless, J. F. (1978). Ridge and related estimation procedures: Theory and practice. Commun. Stat. A Theory Methods 7: 139-164.

Lawless, J. F., and Wang, P. (1976). A simulation study of ridge and other regression estimators. Commun. Stat. A Theory Methods 5: 307-323.

Lawrence, K. D., and Marsh, L. C. (1984). Robust ridge estimation methods for predicting U.S. coal mining fatalities. Commun. Stat. A Theory Methods 13: 139-149.

Lee, T. S., and Campbell, D. B. (1985). Selecting the optimum-k in ridge regression. Commun. Stat. A Theory Methods 14: 1589-1604.

McDonald, G. C., and Galarneau, D. I. (1975). Monte-Carlo evaluation of some ridge-type estimators. J. Am. Stat. Assoc. 70: 407-416.

Montgomery, D. C., and Peck, E. A. (1982). Introduction to Linear Regression Analysis. Wiley, New York.

Narula, S. C., and Wellington, J. F. (1982). The minimum sum of absolute errors regression: A state of the art survey. Int. Stat. Rev. 50: 317-326.

Neter, J., Wasserman, W., and Kutner, M. H. (1985). Applied Linear Statistical Models. Richard D. Irwin, Inc., Homewood, Ill.

Nordberg, L. (1982). A procedure for determination of a good ridge parameter in linear regression. Commun. Stat. B Simul. Comput. 11: 285-309.

Pagel, M. D. (1981). Comment on Hoerl and Kennard's ridge regression simulation methodology. Commun. Stat. A Theory Methods 10: 2361-2367.

Pfaffenberger, R. C., and Dielman, T. E. (1985). A comparison of robust ridge estimators. Proceedings of the American Statistical Association Business and Economic Statistics Section, Las Vegas, Nev., pp. 631-635.

Pfaffenberger, R. C., and Dielman, T. E. (1984). A modified ridge regression estimator using the least absolute value criterion in the multiple linear regression model. Proceedings of the American Institute for Decision Sciences, Toronto, pp. 791-793.

Precht, M., and Rao, P. (1985). An evaluation of biased estimators of regression coefficients—a simulation study. Stat. Hefte 26: 263-285.

Ramsay, J. O. (1977). A comparative study of several robust estimates of slope, intercept, and scale in linear regression. J. Am. Stat. Assoc. 72: 608-615.

Silvapulle, M. J. (1985). An examination of criticisms of ridge regression simulation methodology. Commun. Stat. B Simul. Comput. 14: 829-835.

Sposito, V. A., and Tveite, M. D. (1986). On the estimation of the variance of the median used in L_1 linear inference procedures. Commun. Stat. A Theory Methods 15: 1367-1375.

Vinod, H. D. (1978). A survey of ridge regression and related techniques for improvements over ordinary least squares. Rev. Econ. Stat. 60: 121-131.

Vinod, H. D., and Ullah, A. (1981). Recent Advances in Regression Methods. Marcel Dekker, New York.

Wagner, H. (1959). Linear programming techniques for regression analysis. J. Am. Stat. Assoc. 54: 206-212.

Wichern, D. W., and Churchill, G. A. (1978). A comparison of ridge estimators. Technometrics 20: 301-311.

Yohai, V. J., and Maronna, R. A. (1979). Asymptotic behavior of M-estimators for the linear model. Ann. Stat. 7: 258-268.

14

Composite Earnings Forecasting Efficiency and Executive Composition

JOHN B. GUERARD, Jr. Drexel Burnham Lambert, Chicago, Illinois

ROBERT OCHSNER Hay Management Consultants, Philadelphia, Pennsylvania

1. INTRODUCTION

Recent studies have shown that composite forecasting produces forecasts that are superior to individual forecasts. This chapter extends the existing literature by employing robust regression techniques in composite model building. Security analysts' forecasts may be improved when combined with time series forecasts for a diversified sample of 261 firms with a 1980–1982 postsample estimation period. The mean square error of analyst forecasts may be reduced by combining analyst and univariate time series model forecasts in an ordinary least squares regression model. This reduction is very interesting when one finds that the univariate time series model forecasts do not substantially deviate from those produced by ARIMA (0, 1, 1) processes. Multicollinearity exists between analyst and time series model forecasts and ridge regression techniques are used to estimate composite earnings models. Moreover, the use of robust regression and biased regression techniques produces a 51.3% mean square forecasting reduction from the original analyst forecasts. Executive compensation should be associated with earnings growth because of the linkages among earnings, stock prices, and stock option exercise.

2. REVIEW OF THE LITERATURE

The majority of the literature supports the conclusion that earnings forecasts prepared by security analysts are more accurate than time series model forecasts (Fried and Givoly, 1982; Armstrong, 1983); however, not all of the economic studies have supported the forecasting efficiency of analysts (Cragg and Malkiel, 1968; Elton and Gruber, 1972; Guerard, 1987). The purpose of this study is to develop models combining analyst and time series forecasts to more effectively forecast corporate earnings. The univariate time series model forecasts generally do not violate the ARIMA (0,1,1) process (Ball and Watts, 1979). The majority of researchers have analyzed the annual earnings-generating process and found that a random walk with a first-order moving-average operator best describes the series Albrecht et al., 1977; Watts and Leftwich, 1977). Analyst forecasts may be combined with univariate time series model forecasts in order to produce superior models for estimating earnings. For the sample used in this study, time series earnings forecasts combined with analyst forecasts produce more efficient earnings forecasts (Guerard, 1987), supporting the results of Conroy and Harris (1987).

The purpose of this study is to develop composite models and examine the issue of robust-weighting in estimating composite earnings models. Granger and Ramanathan (1984) propose a method of combining forecasts with no restrictions on the weights. Moreover, a constant term is estimated. The unrestricted weighting scheme of combining forecasts with a constant term produces an unbiased forecast and will produce the lowest estimated mean square error. Multicollinearity exists between analyst forecasts and time series forecasts and leads one to examine the application of ridge regression techniques. One of the findings of the present study is that the estimation of ridge estimators may improve the stability of the regression coefficients.

3. SECURITY ANALYST FORECASTS

The one-year-ahead security analyst forecasts used in this study are those published in the Earnings Forecaster by Standard and Poor's. Analyst forecasts are collected and published in the weekly periodical. Elton et al. (1981) have shown that analyst forecasts are immediately incorporated into the share prices such that one cannot earn an excess return by purchasing securities forecast to have the highest growth prospect; one can profit only by purchasing securities that achieve a higher than expected earnings growth. The purpose of this study is to develop composite earnings models. Better forecasting models may identify potentially high-growth securities overlooked by analysts, or may produce more accurate forecasts of growth rates than those produced by analysts. The identification of such securities could produce high portfolio yields for the investor. The mid-April Earnings Forecaster issues are used because the annual earnings of the previous year are generally known by April.

4. UNIVARIATE TIME SERIES MODEL BUILDING AND FORECASTING

Univariate time series models are estimated for 261 randomly selected firms, during the period from 1959 to 1979. Only 21 observations may seem quite short for time series modeling (Newbold and Granger, 1974); however, the Ljung-Box statistics and residual plots indicate that the models are adequately fitted and the first-difference models are stationary. Moreover, the accounting literature is rich with time series studies using only 25 annual observations (Albrecht et al., 1977).

Annual data on earnings per share are taken from the COMPUSTAT tapes and from Moody's Industrial Manual for the 261 firms. The general form of the models is that of an ARIMA (0,1,1) process, a random walk with drift series. Although many of the models possess higher-order moving-average parameters than the first-order parameter, the estimated model forecasts are so highly correlated with random walk with drift forecasts that one cannot reject the Little (1962) hypothesis of random earnings changes. Logarithmic transformations were applied to all earnings series because of the linear relationship between the series ranges and means in 10-year intervals (Jenkins, 1979).

5. COMPOSITE EARNINGS ESTIMATIONS

The use of ordinary least squares (OLS) in estimating composite earnings models as developed by Granger and Ramanathan (1984) reduces the average estimated mean square regression error relative to the analyst forecasts for the 261 firms for the 1980-1982 postsample period. The estimated composite earnings models are summarized in Table 1 and the 1980-1982 average mean square estimation errors are shown in Table 2. The use of a large sample of firms produced an interesting result for relative estimating efficiency; the average 3-year analyst (SEC) mean square forecasting error of 1.270 is approximately equal to the univariate time series (BJU) mean square forecasting error of 1.277 shown in Table 2. This result supports the early work of Cragg and Malkiel (1968), Elton and Gruber (1972), and Guerard (1987), previously described in analyzing analyst forecasting efficiency.

TABLE 1 Regression Analysis Summary

Year	Regression equation	Dependent variable	Constant	BJU	SEC	MSE	k	R^2
1980	OLS	EPS	-.255	.610	.523	.236		.595
		(t)	(-2.85)	(6.53)	(6.98)			
	Ridge	EPS	.181	.452	.371	.262	.67	
				(12.91)	(13.25)			
	LAD (least absolute deviation)	EPS	-.029	.453	.570			
	Robust	EPS	-.147	.583	.484	.170		
			(-1.92)	(7.30)	(7.56)			
	Ridge (robust)	EPS	.138	.485	.379	.138	.42	
				(15.20)	(14.60)			

1981	OLS	EPS	−.256 (−2.88)	.427 (4.56)	.648 (7.82)	.276		.576
	Ridge	EPS	.219	.363 (12.10)	.378 (13.64)	.316	.83	
	LAD	EPS	−.041	.353	.652			
	Robust	EPS	−.225 (−3.00)	.423 (5.35)	.641 (9.09)	.193		
	Ridge (robust)	EPS	.200	.365 (12.20)	.391 (14.50)	.300	.78	
1982	OLS	EPS	−.271 (−2.26)	.435 (3.49)	.564 (5.05)	.540		.370
	Ridge	EPS	.166	.337 (8.43)	.341 (9.47)	.574	.85	
	LAD	EPS	−.124	.559	.452			
	Robust	EPS	−.258 (−3.02)	.403 (4.51)	.619 (7.64)	.272		
	Ridge (robust)	EPS	−.048	.413 (13.80)	.456 (16.90)	.185	.46	

TABLE 2 Mean Square Forecasting Errors, 1980-1982

BJU	1.277
SEC	1.270
Combined model; OLS	1.040
Ridge	1.139
LAD	1.092
Equal weighted	1.147
Robust	0.729[a]
Ridge (robust)	0.619[a]

[a] Statistically different from security analyst (SEC) forecasts at the 5% level.

The work of Granger and Ramanathan (1984) and Bopp (1985) leads to using OLS analysis to estimate composite earnings models. The cross-sectional composite earnings model is of the form

$$EPS_t = a + b_1 BJU_t + b_2 SEC_t + e_t \tag{1}$$

where

EPS = logarithm of actual earnings per share
BJU = logarithm of univariate time series model forecast
SEC = logarithm of consensus analyst forecast
e_t = randomly distributed error term

Table 2 shows that the OLS composite earnings model-estimated mean square error (MSE) (using the analyst and univariate time series forecast) is 1.040, some 18.1% less than the analyst forecasts. The relatively poor (nondominant) analyst forecasting performance was due to the 1982 analyst forecasts. The regression coefficients on the univariate time series forecast variable in 1980 through 1982 are such that the variable could not be omitted from the regression equation, equation (1). Fried and Givoly (1982) did not find support for the construction of composite earnings models using time series and analyst forecasts whereas Conroy and Harris (1987) found composite modeling effective for forecasting corporate earnings. The principal difference between the

Fried and Givoly study and this study is that Fried and Givoly used a linear correction technique to examine the incremental value of the time series forecast, whereas this study examines the estimation of OLS and biased regression models using both raw variables and avoids the inappropriate application of the incremental value technique (Christie et al., 1984).

Near-multicollinearity exists between analyst forecasts and univariate time series forecasts in the sample period (1959-1979) and post-sample period (1980-1982) and leads one to question the appropriateness of OLS; near-multicollinearity inflates the standard error of the regression coefficients and t-values are biased downward. One would expect that analysts use some variation of a first-order exponential smoothing model [which, of course, approximately equals the ARIMA (0,1,1)]. Thus, one would expect multicollinearity, given that the forecasts are not truly derived from independent sources of information. Fried and Givoly (1982), as noted earlier, found little support for composite model building with time series and analyst forecasts; however, they used a linear correction technique rather than biased regression to reduce multicollinearity. The use of ridge regression techniques on the unstandardized (raw) variables with the Hoerl et al. (1975) iterative procedure to estimate the biasing parameter, k, produces more stable regression coefficients, particularly in 1982 (see Table 1). The regression coefficients on analyst and time series forecasts tend to equality in the ridge regression, as one would have expected, given the approximately equal standard deviations of the variables (Bates and Granger, 1969). However, the more stable regression coefficients of the raw (unstandardized) ridge regression do not estimate earnings as well as the OLS regressions (the 1980-1982 average mean square estimation error is 1.139 with ridge regression).

6. ROBUST REGRESSION AND COMPOSITE MODEL BUILDING

The ordinary least squares regression and ridge regression analyses produce 5 to 10 outliers (observations not within two standard deviations of the regression lines) in each annual regression; given a normal distribution of 261 observations, one would have expected at least 12 or 13 observations to lie outside the confidence intervals. One could use the Beaton-Tukey (1974) biweight (robust) procedure for iteratively reweighting the regressions. Large residuals lead to very small observation weights. The biweight function is

$$w_i = \left[1 - \left(\frac{r}{B}\right)^2\right]^2 \quad \text{if } |r| \leq B$$

$$= 0 \quad \text{otherwise}$$

where

r = absolute value (residual/standard deviation of error)
B = a tuning constant, 4.685

The use of the iteratively reweighted least squares regression and ridge regression techniques substantially reduces the 1980-1982 mean square forecast errors (0.729 and 0.619, respectively, as shown in Table 2). Security analyst forecast errors can be substantially (and statistically significantly) reduced by 42.6 and 51.3%, respectively, with the use of weighted least squares and weighted ridge regression techniques.

The regression coefficients on the time series and analyst forecast variables for the ridge regression on the robust-weighted data in 1981 and 1982 are approximately equal; one would have expected this given the respective (approximately equal) standard deviations of the variables (Bates and Granger, 1969). The ridge-estimates coefficients are slightly better (not statistically different) forecasting coefficients than are obtained by using equal variable weights; the average mean square error is 1.149 with equal weighting. The superiority of the estimated weighting scheme has been advanced by Winkler and Makridakis (1983).

7. FORECASTING, NOT ESTIMATING, COMPOSITE EARNINGS MODELS

The composite models developed in the previous section were constructed by regressing time series forecasts and time series model forecasts on earnings per share. The regression models involved contemporaneous variables and "nonforecasting" models were estimated. However, the use of composite earnings models at time t, based on estimated weights at time t - 1, improves on security analyst forecasts (see Table 3).

The composite earnings models are quite useful for forecasting, particularly in 1982. The application of robust regression techniques is statistically significant in reducing analyst forecasting errors. Moreover, the calculation of firm (observation) weights proves useful in forecasting corporate earnings.

TABLE 3 Mean Square Forecasting Errors
Forecasting Years

Source (year of weights)	1981	1982
SEC (t)	.328	.656
Combined model, OLS (t)	.273	.534
Ridge (t)	.312	.568
Robust (t)	.437	.167
Ridge (robust) (t)	.391	.183
OLS (t - 1)	.283	.550
Robust (t - 1)	.273	.173[a]
Ridge (robust) (t - 1)	.289	.213[a]

[a] Statistically significantly different from analyst forecasts at 5% level.

8. COMPOSITE MODELS AND INFORMATION COEFFICIENTS

The application of ordinary least squares and ridge regression analyses on the robust-weighted data produced the lowest mean square forecasting errors among the composite forecasting models. The reweighted regression techniques also aided in the identification of the relative ranking of the firms' earning growth rates. The information coefficient (IC) measures the efficiency of identifying the relationship between actual and

TABLE 4 Information Coefficients

	1980	1981	1982
BJU	.347	.386	.391
SEC	.468	.384	.364
Combined model, OLS	.353	.280	.312
Ridge	.482	.455	.507
LAD	.525	.442	.472
Robust	.448	.299	.355
Ridge (robust)	.493	.464	.532

forecasted rankings among the sources of information. One would want to maximize the information coefficient b where

actual ranking = a + b(forecasted ranking)

The information coefficients shown in Table 4 show the efficiency of the composite ordinary least squares, minimum absolute deviations, ordinary least squares regression on the robust-weighted data (Robust), and ridge regression on the robust-weighted data [Ridge (Robust)] relative to the analyst ICs. The reweighted regression techniques would substantially aid analysts in forecasting 1981-1982 earnings.

9. IMPLICATIONS FOR EXECUTIVE COMPENSATION

Firms that experience the highest actual earnings growth should produce the highest stock price appreciation (Elton et al., 1981). Thus, the highest growth firms, generating large stock price appreciation, should produce the highest levels of stock option appreciation. The 261-firm sample is used to test the hypothesis that the (actual) highest growth firms produce greater exercise of stock options for the firms' executives. Executive salary and bonus compensation (SB) and stock option compensation (SO) data may be found in Forbes. The 261-firm sample is divided into (1) firms producing the highest earnings growth, when one uses the highest quartile (66 firms) of earnings growth, and (2) the remaining firms (195 firms). One would expect that the highest growth firms produce the largest exercise of executive stock options; however, the dollar amounts of stock option compensation and ratios of stock option compensation to executive salary and bonus levels (CR) of the two samples are not statistically different. One would not expect these results. There may be a problem in the comparison of sample means and variances. That is, in many cases firms allow the exercise of executive stock options within a 90-day "window" of board meetings. If a firm experiences earnings growth and stock price appreciation in late 1980, for example, some of its executives may not exercise the stock options until 1981. Thus, one might expect to find a one-year lag in the exercise of stock options. It appears that the one-year lag scenario is appropriate for the 261-firm sample in 1980; see Table 5 for complete results. Many compensation plans allow management up to five years to exercise options and statistically modeling such a process is difficult at best. Further research is necessary to examine the timing of earnings, stock prices, and exercise of stock options.

TABLE 5 Executive Compensation Sample Statistics

Sample		SB ($1000)			SO ($1000)		
		1980	1981	1982	1980	1981	1982
Highest growth	x	460.9	544.7	515.3	166.0	207.1	184.3
	σ	223.0	252.1	230.3	316.0	269.7	333.2
Nonhighest growth	x	462.3	522.4	544.6	194.4	240.3	217.7
	σ	194.7	219.8	228.4	393.4	453.8	339.5

Sample		CR_t			CR_{t-1}	
		1980	1981	1982	1980	1981
Highest growth	x	.372	.354	.320	.769[a]	.455
	σ	.724	.448	.608	1.262	.595
Nonhighest growth	x	.381	.441	.367	.390	.469
	σ	.751	.829	.541	.762	.871

[a] Statistically different from nonhighest growth sample at 5% level.

10. SUMMARY AND CONCLUSIONS

Composite earnings per share models may be developed using analysts forecasts and univariate time series forecasts. The OLS composite earnings models substantially reduce the mean square forecasting error present in the analyst forecasts despite the fact that the univariate time series models are not significantly different from random walks with drift formulations. The use of biased regression techniques aids in earnings forecasting once the analyst and time series forecast variables are weighted in a robust regression-formulated scheme. The estimation of composite earnings models may be an avenue of potential profit; univariate time series models complemented the analyst forecasts in this study. Furthermore, the results of this study indicate that analysts may not necessarily use all information available at the time of their forecasts.

Moreover, executive compensation may be associated with earnings growth, although the timing of the association may be difficult to validate statistically.

ACKNOWLEDGMENTS

The author appreciates the assistance of Carl Ratner of Standard and Poor's Corporation. The author acknowledges the comments of Professors J. S. Armstrong, R. T. Clemen, C. W. J. Granger, S. Makridakis, J. P. C. Kleijnen, and M. S. Rozeff on previous versions of this study. An earlier version was presented at the American Accounting Association meeting, New York City, August 1986. Any errors remaining are the responsibility of the author. Statistical analysis was performed with the aid of the SAS programs. Access to SAS was provided by the Hay Group and Air Products and Chemicals, Inc.

REFERENCES

Albrecht, S. W., Lookabill, L. L., and McKeown, J. C. (1977). The time-series properties of annual earnings. J. Accounting Res. 15: 226.

Armstrong, J. S. (1983). Relative accuracy of judgmental and extrapolating methods in forecasting annual earnings. J. Forecasting 2: 437.

Ball, R., and Watts, R. (1979). Some additional evidence on survival biases. J. Finance 34: 197.

Bates, J. M., and Granger, C. W. J. (1969). The combination of forecasts. Operational Res. Q. 20: 451.

Beaton, A. E., and Tukey, J. W. (1974). The fitting of power series, meaning polynomials, illustrated on band-spectroscopic data. Technometrics 16: 147.

Bopp, A. E. (1985). On combining forecasts: Some extensions and results. Manage. Sci. 31: 1492.

Box, G. E. P., and Jenkins, G. M. (1970). Time Series Analysis: Forecasting and Control. Holden Day, San Francisco.

Brown, L. D., and Rozeff, M. S. (1978). The superiority of analyst forecasts as measures of expectations: Evidence from earnings. J. Finance 33: 1.

Christie, A. A., Kennelley, M. D., King, J. W., and Schaefer, T. F. (1984). Testing for incremental information content in the presence of multi-collinearity. J. Accounting Econ. 6: 205.

Conroy, R., and Harris, R. S. (1987). Consensus forecasts of corporate earnings: Analysts' forecasts and time series methods. Manage. Sci. 33: 725.

Cragg, J. G., and Malkiel, B. (1968). The consensus and accuracy of some predictions of the growth of corporate earnings. J. Finance 23: 67.

Clemen, R. T., and Winkler, R. L. (1986). Combining economic forecasts. J. Econ. Business Stat. 4: 39.

Elton, E. J., and Gruber, M. J. (1972). Earnings estimation and the accuracy of expectational data. Manage. Sci. 18: 409.

Elton, E. J., Gruber, M. J., and Gultekin, M. (1981). Expectations and share prices. Manage. Sci. 27: 875.

Farrell, J. (1983). Guide to Portfolio Management. McGraw-Hill, New York.

Fried, D., and Givoly, D. (1982). Financial analysts' forecasts of earnings: A better surrogate for market expectations. J. Accounting Econ. 4: 85.

Forbes. How Much Does the Boss Make? June 8, 1981: 114.

Forbes. Who Gets the Most Pay. June 7, 1982: 74.

Forbes. The $50 Million Man. June 6, 1983: 126.

Granger, C. W. J., and Newbold, P. (1977). Forecasting Economic Time Series. Academic Press, New York.

Granger, C. W. J., and Ramanathan, R. (1984). Improved methods of combining forecasts. J. Forecasting 3: 197.

Guerard, J. B. (1987). Linear constraints, robust-weighting and efficient composite modeling. J. Forecasting 6: 193.

Guerard, J. B., and Beidleman, C. R. (1986). A new look at forecasting annual earnings in the U.S.A. Eur. J. Operational Res. 23: 288.

Guerard, J. B., and Beidleman, C. R. (1986). Composite forecasting of annual earnings: An application of biased regression techniques. J. Stat. Comput. Simul. 1, 23: 1.

Hoerl, A. E., and Kennard, R. W. (1970). Ridge regression: Biased estimation for nonorthogonal problems. Technometrics 12: 55, 69.

Hoerl, A. E., Kennard, R. W., and Baldwin, K. F. (1975). Ridge regression: Some simulations. Commun. Stat. 4: 105.

Jenkins, G. M. (1979). Practical experiences with modelling and forecasting time series. Forecasting, O. Anderson, ed. North-Holland, Amsterdam.

Little, I. M. D. (1962). Higgledy piggledy growth. Oxford Univ. Inst. Stat. 24: 387.

Malkiel, B. (1981). A Random Walk Down Wall Street, 2d ed. Norton, New York.

Mason, R. L., Gunst, R. F., and Webster, J. T. (1975). Regression analysis and problems of multi-collinearity. Commun. Stat. 4: 277.

Montgomery, D. C., and Peck, E. A. (1982). Introduction to Linear Regression Analysis. Wiley, New York.

Newbold, P., and Granger, C. W. J. (1974). Experience with forecasting univariate time series and the combination of forecasts. J. R. Stat. Soc. 137: 131.

Vinod, H. D. (1978). A survey of ridge regression and related techniques for improvements over ordinary least squares. Rev. Econ. Stat. 58: 121.

Vinod, H. D., and Ullah, A. (1981). Recent Advances in Regression Methods. Marcel Dekker, New York.

Watts, R. L., and Leftwich, R. W. (1977). The time series of annual accounting earnings. J. Accounting Res. 15: 254.

Webster, J. T., Gunst, R. F., and Mason, R. L. (1974). Latent root regression analysis. Technometrics 16: 513.

Winkler, R. L., and Makridakis, S. (1983). The combination of forecasts. J. R. Stat. Soc. 146: 150.

Index

Algorithm:
- complete search, 116-117
- dual, 95-103
- for finding multiple minima, 116-120
- making use of the design, 109
- primal, 96-103
- treating k as a free parameter, 120
- using residuals, 118-119

Autocorrelation, 195-196

Bayesian, 35
- posterior mean, 11-20

Branch-and-bound algorithm, 93-103
- LIFO branching, 93-103

Dual problem, 24-25

Forecasting:
- accuracy, 195-198
- efficiency, 269-282
- univariate time series model, 269-282

Gompertz model, 65-86

Input/output models, 181-194

L_1 estimators, 25-27
- asymptotic distribution, 48-52
- computational procedures, 27-34
- equivalence of L_1 and R-estimators and nonlinear

[L_1 estimators]
programming, 59-86
least absolute error, 176-180
properties, 25-27, 32-34
robust time series, 173-180
subroutine SIMPL, 31

L_p estimators, 23-55, 130-135
Median Polish, 40-48
optimal, 36
unbiased, 34-36

LAV estimators, 71-86, 89-103, 243-270

Least Squares:
iteratively reweighted, 114-116
maximum likelihood, 192-194

Logistic model, 71-86

Maximum likelihood, 176-180, 192-194

Mergers, 213-223

M-estimators, 11-20, 105-128
redescending 106, 243-270

Multicollinearity, 243-270

Nonnormal error distributions, 243-270

Ordinary least squares, 71-86, 215, 243-282

Outliers:
alternative methods of dealing with, 225-239
and multicollinearity, 243-270
detection, 195-211
time series, 198-211
multiple multivariate, 89-103
robust regression, 190-192

Posterior mean, 5-11

Primal problem, 24-25

Regression:
balanced one-way model, 144-151
functions, 108-113
latent root, 216-223
general mixed model, 151-153
robust estimation, 154-168
multicollinearity and outliers, 243-270
residuals, 143-172
ridge regression, 243-270
WRID (weighted ridge), 243-270
simple linear minimum sum of absolute error regression, 269-282
robust, 243-282
estimation, 3-5
regression, 59-86, 129-141, 190-194
time series analysis, an L_1 approach, 173-180
weighting scheme, 218-219

Regression scheme:
Anscombe, 77-86
Hill Holland, 77-86
Huber, 77-86
Ramsey, 77-86
unweighted, 77-86

Sigmoidal growth models, 65

Simple location model, 8-11

Theory of complementary slackness, 24-25

Time series analysis, 173-180
univariate, 269-282

Tukey's biweight, 20, 106-108

Variance component models, 143-172

Weighting curves, 20

Weighting structure, 5-11

Printed in the United States
by Baker & Taylor Publisher Services